ARZAMAS-16

Russian Memoirs Series No. 3

Also by Michael Pursglove

DEAF CHILDREN AND BILINGUAL EDUCATION

N. V. GOGOL'S *DIARY OF A MADMAN* (EDITOR)

THE NEW RUSSIA

N. V. GOGOL'S *NEVSKY PROSPECT* (EDITOR)

L. N. TOLSTOI'S *SEBASTOPOL IN DECEMBER/ SEBASTOPOL IN MAY* (EDITOR)

L. N. TOLSTOI'S *CHILDHOOD* (EDITOR)

D. V. GRIGOROVICH'S *ANTON; THE PEASANT* (TRANSLATOR)

ARZAMAS-16

Soviet Scientists in the
Nuclear Age: a Memoir

by
Veniamin Tsukerman

and
Zinaida Azarkh

Translated by Timothy Sergay

Introduced and edited by
Michael Pursglove

Bramcote Press
Nottingham

Original text in Russian by V.A. Tsukerman and Z.M. Azarkh in book form entitled *Люди и взрывы*
© РФЯЦ–ВНИИЭФ, 1994, ISBN 5-85165-058-3. Права защищены. Данное издание (или его части) не могут быть воспроизведены, занесены в запоминающие и воспроизводящие системы или переданы в какой-либо форме или каким-то образом без письменного разрешения.
Переводчик подтверждает аутентичность перевода.
The translator confirms that the translation is authentic.
The above statements are made for reasons of copyright.

Translation © Richard Rhodes 1999
All additional material © Michael Pursglove 1999

Acknowledgement is made to the Alfred P. Sloan Foundation, which supported in part the English translation of the text, and to the School of Modern Languages and the Department of Russian, both of the University of Exeter, for financial assistance in the publication of this book.

All rights reserved. No reproduction, copy or transmission of this publication may be made without written permission.

This edition of the authentic translation first published 1999 by **BRAMCOTE PRESS, Nottingham, England.**

Printed in Great Britain by
Antony Rowe Ltd, Chippenham, Wiltshire

Cover design by MAM
4 Russell Place, Nottingham

British Library Cataloguing in Publication Data.
A catalogue record for this book is available from the British Library.

ISBN 1 900405 04 0

Contents

List of Illustrations	*vii*
Introduction	*ix*
Behind the Iron Shutters of Secrecy	*xxiii*
Foreword	*xxix*
Prologue	*xxxiii*
PART 1: Vitebsk-Moscow-Kazan-Moscow	**1**
My Vitebsk	1
Friends of the 1930s	5
First inventions	8
Zina Azarkh	12
My eyes	17
The first day of the war	21
The first month	23
Kazan	24
The bottle-launcher	27
Lida	29
14 July 1942	32
X-ray of an explosion	34
Work and life without electricity	38
Moscow 1944-47	41
Mother, don't die	41
Kapitsa's seminar	43
The first award	44
Visit to Leningrad	44
August 1945	45
Talks with Khariton	47
PART 2: On Another Meridian	**49**
The beginning of a new life	49
The Mirror	58
Safety procedures	59

Flies	63
The library	63
Romance and life	65
The dispute	71
Critical stages	73
A raw recruit	74
The first test	75
PART 3: The ones who began	**79**
V.A. Aleksandrovich	79
L.V. Al'tshuler	83
A.K. Bessarabenko	85
A.A. Brish	87
V.A. Davidenko	90
M.V. Dmitriev	95
S.B. Kormer	98
V.V. Sof'ina	102
D.M. Tarasov	104
P.M. Tochilovskii	106
M.A. Kanunov	107
A.A. Zhuravlev and I.I. Ignat'ev	110
PART 4: Titans of the Human Spirit	**111**
I.V. Kurchatov	111
Iu.B. Khariton	121
I.B. Zel'dovich	140
I.E. Tamm	145
A.D. Sakharov	149
P.M. Zernov	158
B.G. Muzrukov	163
Epilogue	*169*
Afterword	*170*
Glossary of Russian Terms	*171*
Bibliography	*172*
Who's Who and Index of Personal Names	*174*

Illustrations

Between pages 78 and 79

The authors in 1933
The authors in 1978
Tsukerman and N. K. Reshetskaia in the X-ray laboratory, 1941
X-ray photograph of a bullet passing through a light bulb taken just before the glass has shattered, 1945
The Institute's first research laboratory was accommodated in this building in 1947
High-tension 500 kv. impulse radiography apparatus, 1946
The first purpose-built laboratory, constructed in 1948
M. A. Manakova assembles the charge for X-ray experiments
View of the town of Sarov in 1990
V.A. Aleksandrovich
L.V. Al'tshuler
A.K. Bessarabenko
A.A. Brish
V.A. Davidenko
M.V. Dmitriev
S.B. Kormer
V.V. Sof'ina
D.M. Tarasov
P.M. Tochilovskii
M.A. Kanunov
I.V. Kurchatov
Iu.B. Khariton
Ia.B. Zel'dovich
I.E. Tamm
A.D. Sakharov
P.M. Zernov
B.G. Muzrukov

Russian words and expressions not explained in the text of this book are defined in the *Glossary* on page 171.

Persons mentioned in the Memoir and in the introductory sections are identified in the *Who's Who and Index* at the end of the volume.

Introduction

THE TRAIN which leaves Moscow's Kazan station every evening is shown on the indicator boards as being bound for Pervomaisk. This nondescript town, due south of Gor'kii/Nizhnii Novgorod, is an overnight journey away from Moscow. Some of the carriages, including, frequently, one which was formerly the personal carriage of Tsar Nicholas II, are marked 'Bereshcheno', a village in the Mordovan Autonomous Republic. The real destination of these carriages is not, however, Bereshcheno, but a town which, in Soviet times, was never mentioned in print and never indicated on any map: Sarov, or, to give it its code name, Arzamas-16, home of the top-secret nuclear research facility where the Soviet nuclear bomb was designed and built. The Tsar's carriage was for the use of the redoubtable Iulii Khariton, Scientific Director of Arzamas-16 from 1947 until 1992 and Emeritus Scientific Director until his death in 1996.

Passengers making for Arzamas-16 by train have to have full security clearance, and buy their tickets through the organs of State Security. The same is true of those who make the one-hour flight from Moscow's Bykovo airport. The plane's destination is not indicated, and all check-in formalities for the 'Special Flight' *(spetsreis)* are done by security officials. At one time the length of the flight was a State secret.

The town of Sarov lies in the Temnikov region of the Mordovan Autonomous Socialist Soviet Republic, a region once inhabited by the Finnic-speaking Mordva people, but now fully russified. Arzamas-16/Sarov is situated on the River Sarovka some five hundred kilometres east of Moscow by the most direct route. The nearest large town is Nizhnii Novgorod, some two hundred kilometres distant to the north. It has a population of 80,000 and is surrounded by a barbed wire fence fifty-six kilometres long, patrolled by two thousand Interior Ministry troops. Before the arrival of the nuclear scientists the official status of Sarov was that of a 'settlement' *(poselok)*, rather less than a town, with a population of between two and three thousand. Sarov was famous for its monastery, founded in 1679, which had once numbered among its monks the second most venerated saint in the Russian Orthodox Church after St Sergius of Radonezh, St Seraphim of Sarov. The name Arzamas-16, chosen by Iulii Khariton in 1945, was one of several employed to disguise the existence of the Soviet Union's nuclear weapons research facility: Base-112; KB-11; Kremlev; Arzamas-75; Installation (ob"ekt) No. 550; Moskva; Centre-300, the Volga Office *(Privolzhskaia kontora)*. Unofficially, in a backhanded compliment to the American success in producing a nuclear bomb, it was known as Los Arzamas, and sometimes simply as Atomgrad. The name finally adopted is, of course, deliberately misleading: the real town of Arzamas, where Lev Tolstoi had a vision of death in 1869, is some sixty kilometres to the north of Arzamas-16. In the same way the nuclear weapons laboratory known as Cheliabinsk-70, established by scientists from Arzamas-16 in 1955, is actually situated not in Cheliabinsk but at Kasli, some eighty kilometres to the north. In his memoirs, Andrei Sakharov, who worked at Arzamas-16 between 1950 and 1968, refers to the place as 'The Installation' and refuses to write about his work there in detail. He does, however, memorably describe the place as 'a curious symbiosis between an ultramodern scientific research institute, with its experimental workshops and testing grounds—and a huge labour camp.' Despite this plethora of code names, the scientists always referred to their place of work as Sarov. The author of these lines, who first met Veniamin Tsukerman and Zinaida Azarkh in Moscow in September 1989, was not allowed until 1992 to know where they worked. Thirteen years earlier, in a letter to the writer Iurii Nagibin, Tsukerman had emphasized the need for total secrecy. Nagibin had published a short story entitled 'The Spring that Fell Silent' ('Zamolchavshaia vesna') which included thinly disguised portraits of Tsukerman, his wife, his daughter and grand-daughter. No names were given, and Tsukerman was portrayed as a mathematician rather than a physicist. However, many of the details were unmistakable: Tsukerman's blindness, his daughter's deafness, the death of his son, even, as Tsukerman pointed out in the letter, the colour of his wife's eyes. The story, which tells how a man gradually becomes deaf, was due to be republished in the journal of VOG, the All-Russian Society of the Deaf, with an intro-

Introduction

ductory commentary by the author. Tsukerman makes an urgent request of the author:

> Do not under any circumstances reveal our surnames, or even the existence of our unique family. When your story was first published we had a number of micro-unpleasantnesses along these lines. You are, of course, free to write what you like about the prototype of your hero, Sergeev...But, for obvious reasons, you must not add anything at all to what has already been published.

So great was the secrecy that some employees at Arzamas-16 were unaware that they were working on a hydrogen bomb. A former director of Arzamas-16, E.A. Negin, tells a possibly apocryphal anecdote to the effect that, while leaning against the bomb itself, an employee expressed amazement that Malenkov should have announced that the Soviet Union had a hydrogen bomb and could only conclude that it had been made at some other centre. The journalist Oleg Moroz described the régime of secrecy as 'a way of life, which determined the way people behaved and thought, and even their spiritual condition.' Unintentional breaches of security, such as that described in Part 2 in the section 'Romance and Life', could have dire consequences and were the cause of at least two suicides. One suicide was that of the head of security at Arzamas-16, held responsible for the fact that, in an institution in which no typewriters could be held privately, an officially sanctioned typist had failed to record the destruction of a piece of paper. The typist was jailed for a year. In another incident, recorded by Sakharov, a scientist Boris Smagin lost a secret piece of equipment down the toilet. After a search through frozen sewage by the NKVD, the missing piece was found. Smagin was dismissed but forbidden to leave Arzamas-16.

Arzamas-16 came into being officially with a governmental decree dated 21 June 1946 (which referred to the place as KB-11). Evidently the workforce was insufficient, because on 13 February 1947 P.M. Zernov, the first director of Arzamas-16, complained to Moscow that no building had yet been completed in the town. The following day the town was declared a 'strict régime zone' *(rezhimnaia zona)*, operated directly from Moscow. In July of the same year a hundred and eight local families were moved out of the 'zone'.

Initially facilities at Arzamas-16 were minimal. Food, including the inevitable continental sausage, had to be brought in from

Moscow, and all laundry had to be sent out to the capital. By 1991 there were eleven such places in the Soviet Union, known collectively as the 'White Archipelago', to distinguish them from the 'Gulag Archipelago' of labour camps. The total population of the White Archipelago was some 700,000, of whom between ten and fifteen thousand were privy to secret information. This figure included two to three thousand top scientists. In bureaucratic jargon each settlement was known as a ZATO, Russian initials for 'Closed Administrative and Territorial Formation'. Everything about these places was shrouded in secrecy. For example the Ministry which ran them was known first as the 'First Main Directorate of the Council of Ministers of the USSR' and, from 1953 onwards, as the Ministry of Medium Machine Building *(Ministerstvo srednego mashinostroeniia)*, usually shortened to Sredmash. The composition of the Special Committee which, under the control of the Central Committee of the Party, set up Arzamas-16 (Malenkov, N. Voznesenskii, Beria, Zaveniagin, M. Pervukhin, Ioffe, Kapitsa, Kurchatov, and V. Makhnev) was, of course, secret, as were its deliberations. Various organizations set up to service the new weapons centre were known only by code names: Research Institute 6 developed detonators; Research Institute 304 worked on high-grade explosive. The research facilities at Arzamas-16 were dubbed 'the Laboratory of Measuring Instruments', and the ultimate product of these facilities—the atomic bomb and later the thermonuclear, or hydrogen, bomb—was referred to as 'the manufacture' *(izdelie)*. Different installations within Arzamas-16 were referred to by number: thus the aerodrome was Area No. 5, the explosives stores were Area No. 4, and so on. To this place came the cream of the Russian scientific intelligentsia. One of them, Iurii Trutnev, speaking in 1998 to Mark Franchetti, the first British journalist to be allowed entry to Arzamas-16, recalled how he was summoned to the town in 1951: 'I received strict instructions to go to the airport and board a plane with five other scientists. We had no idea where we were being taken.'

Within Arzamas-16 itself secrecy prevailed from the outset. One of the first newcomers was a demobilized member of the fourth Kantemir Tank Corps, V.B. Zlatoustovskii:

> At the beginning of 1946 a group of soldiers who had been prematurely demobilized (thirty-five to forty of them) arrived in Sarov to work. They formed the basis of the workforce of

Introduction

Factory No. 1...On the first of December it was pouring with rain. On that day we learned that we had to fly somewhere. Where to, why—not a word. We waited until four o'clock, then set off by train from the Kazan station. We stopped at Arzamas-2. Outside it was deep midwinter. We drove in cars through the darkness, passing through several hamlets en route. Finally we passed through several 'zones', each fenced off from the next and illuminated along the perimeter, like in prison camps, and arrived at a monastery. After a good meal, we went to our quarters in a settlement consisting of ten to twelve uniform houses. In the morning we were greeted by N.A. Petrov and the deputy director in charge of domestic affairs Kh.A. Kostan'ian.

Petrov chatted with each man and asked what work they had done before the war. He told us we would be working in a defence establishment. The deputy director of the factory F.I. Orlov took us to the factory. The factory was rather like a building site on the White Sea Canal: all around there was wire, watchtowers, guards, dogs...All four or five buildings of the factory were being reconstructed. They had no equipment and no floors. The workforce consisted of about sixty men, including ourselves who had just arrived (outdoor and indoor repair work was done, for the most part, by prisoners). We were introduced to the future heads of each workshop...

Then we were given instructions about correspondence. We were given a special address; Glavgorstroi, Volga Factory, PO Box 214.

The passports of the scientists recruited to work at Arzamas-16 gave non-existent Moscow addresses as their place of residence. Two staff members at Arzamas-16, Leonid Ognev and Vladislav Mokhov, asked by the journalist Vladimir Gubarev whether this secrecy influenced their psychology, or their lives, conceded that it had been difficult in the first year. Mokhov said he was not unduly affected, but that many were. His mother, for example, on a visit to her son at Arzamas-16, had once remarked: 'It's difficult to breathe here.' Khariton himself did not reveal the full extent of his involvement in Arzamas-16 until December 1992, although he had been named as the 'father of the atom bomb' in the crusading newspaper *Argumenty i fakty* in October 1989. This is hardly surprising given Khariton's own nature, which he himself described as 'taciturn and uncommunicative' *(molchalivyi i nerazgovorchivyi)*. Even in August 1993, when asked about his role in the creation of the first atomic bomb, he twice told Vladimir Gubarev that he 'did not have the right to say anything.' In the same interview he admitted that while the demands of secrecy did not oppress him, they did adversely affect other people.

The official code words surrounding the nuclear project begat further, more informal, code words. The nuclear scientists (Russian: *iadershchiki*) working on the 'manufacture' *(izdelie)* were termed 'manufacturers' *(izdel'shchiki)*. Other names for them included 'weapon-makers' *(oruzheiniki)*, 'medium machine builders' *(sredmashevtsy)*, 'blind hawks' *(slepye iastreby)*, or simply 'boffins' *(nauchniki)*.

The mere fact of the publication of this book was a sign that Gorbachev's policy of *glasnost* (openness) was being applied to perhaps the most carefully hidden part of the Soviet system. It originally appeared in the September, October, and November editions of the journal *Zvezda* in 1990, having been completed in 1988. This journal version included an introductory essay by Academician Gol'danskii in which he describes this book as 'an open acknowledgment' of the work of Arzamas-16, and, he implies, a long-overdue one at that. The book edition, published in Arzamas-16, appeared in 1994, without Gol'danskii's essay, which we have restored in this English translation. An extract from the section entitled 'Zina Azarkh' in Part 1 appeared in the journal of the All-Russian Federation of the Deaf in August 1995 (*V edinom stroiu*, 8, 1995, 18-19).

The original Russian title *People and Explosions (Liudi i vzryvy)* was probably influenced by the title of Boris Pasternak's autobiographical prose piece *People and Positions (Liudi i polozheniia*, 1957), a piece perhaps better known by its subtitle *Essay in Autobiography (Avtobiograficheskii ocherk)*. The literary reference is noteworthy. Pasternak himself, an author hounded by the Soviet authorities after the publication of his novel *Dr Zhivago* in 1957, is quoted again at the beginning of the section on Iulii Khariton (Part 4). Khariton was a man steeped in literature, as were many of the scientists at Arzamas-16. Lev Al'tshuler, Tsukerman's oldest friend and a man who can cite swathes of Russian poetry by heart, testifies to the popularity of the poetry of Nikolai Gumilev, executed by the Bolsheviks in 1921, among the scientists at Arzamas-16. Tsukerman himself was, as we are told in the section on Muzrukov in Part 4, responsible for establishing the cultural centre in the town which today bears the

Introduction

name of 'Scientists' House' *(Dom uchenogo)*, and his voluminous correspondence includes letters to his contemporaries in the world of literature and the arts, such as the novelist Konstantin Simonov, the short-story writers Iurii Nagibin and Izrail Metter, the playwright Viktor Rozov and the widow of the film director Grigorii Kozintsev. Typically, the letter to Nagibin, quoted above, also contains a suggestion concerning the ending of the story 'The Spring that Fell Silent' which is motivated by artistic, rather than security, considerations.

As Tsukerman explains in his preface, the book, though written in the first person, was a collaborative venture between himself and his wife, Zinaida Azarkh. Despite his blindness and despite a heart attack in 1971, Tsukerman remained in robust health until almost the end of his life. Only in the 1990s did his health decline to the point where the task of editing the text for book publication fell on his wife.

The fall of communism has brought changes to Arzamas-16. The nomenclature has changed: the ministry responsible, with its huge building on Bol'shaia Ordynka Street, is now known as the Ministry of Atomic Energy, Minatom for short, and the name Arzamas-16 has, at the behest of the inhabitants, reverted to Sarov. The nuclear weapons facility is now known as Nuclear Centre, the initials of which are tacked on to the front of the neutral sounding VNIIEF (All-Russian Institute of Experimental Physics). Books and newspaper articles have been published dealing with Arzamas-16 and its personnel, a US television series and an authoritative book have been devoted to the subject, and the BBC broadcast a documentary in 1998 on Arzamas-16 in a series entitled 'Science at War'. A large road sign now directs the motorist off the main road from Murom and towards Sarov. There are direct e-mail and phone links to the town, (although direct phone calls by individuals from the town to Western countries still appear to be impossible). It was only in the 1970s that the inhabitants could make any telephone calls at all out of the town. In 1995 the newspaper *Izvestia* reported that the continued restrictions on the inhabitants of Arzamas-16/Sarov had been challenged in the courts as an infringement of their civil rights. However, although the judge ruled that such an infringement had taken place, Minatom and the FSB (successor to the KGB security apparatus) were appealing against the decision.

Meanwhile the inhabitants of the town have found benefits in living in a 'closed' town, since they are thus shielded from the steep increase in levels of crime which have bedevilled the inhabitants of 'open' towns. The town has a flourishing cultural life, out of all proportion with its size. Among artists appearing in 'Scientists' House' in 1998 were Mstislav Rostropovich and Galina Vishnevskaia and the Borodin String Quartet.

Some things, however, have not changed. The telephone numbers of private subscribers in the town are still given as Gor'kii numbers; only recently could mail be sent directly to addresses in the town—previously mail had to be sent to a special Moscow address; only direct relatives of scientists have access to the town, relatives by marriage being obliged to the security authorities for each separate visit.

Since the fall of communism and the subsequent crises in the Russian economy, safety has become a major concern at Arzamas-16. Even in Soviet times, when funding was not a problem, there were accidents. Two such are referred to: to M.V. Dmitriev (see Part 3), on which subject the book is reticent, and to Khariton himself (see Part 4), on which subject the book is more forthcoming. The most recent accident appears to have been in June 1997, when Aleksandr Zakharov died after receiving a high dose of radiation. Zakharov was said to be the first Russian to die in a nuclear accident since the Chernobyl disaster of April 1986. However, there had been a spate of non-fatal accidents, three in the Cheliabinsk complex alone in 1993. Security, too, became an issue, when two men were arrested in West Germany carrying amounts of uranium they had apparently obtained from Arzamas-16, presumably with 'inside' help, and which they intended to sell. The incident illustrates the size of the financial problems besetting Arzamas-16, problems which have intensified with the currency collapse of August 1998. Many employees of the laboratory have not been paid for months, and their representatives, together with representatives of workers in the nuclear industry, held a mass demonstration in Moscow in September 1998. The Americans have been providing funding since 1994, in the hope of stemming any 'brain-drain' of nuclear scientists to hostile countries, and in October 1998 announced a further aid package.

The authors of this memoir do not, of course, purport to give a complete history of Arzamas-16. There are, nevertheless, some omissions which mark this work out as a book of the Gorbachev

Introduction

period, when progress towards complete freedom of speech, though dramatic, was constrained by ideology and political pragmatism. Tsukerman does not, for example, name the location of the first Soviet nuclear test on 29 August 1949, although he devotes a section of his book (in Part 2) to it. In fact, as David Holloway tells us, the test was carried out in Kazakhstan, at a site some seventy kilometres to the south of the town code-named Semipalatinsk-21, itself situated about a hundred and forty kilometres north-west of the city of Semipalatinsk. Not only details, but whole issues, are passed over in silence. One such issue is that of anti-Semitism. The word 'Jew' appears only once, even though the authors are both Jewish by race, though not by religion, as were Khariton, Zel'dovich, Al'tshuler, Vannikov, and many others. Sakharov reports that in some circles the sister institution of Arzamas-16, Cheliabinsk-70, was known as 'Egypt', the clear implication being that Arzamas-16 was 'Israel'. Sakharov himself was not Jewish, but, when, in 1968, he attacked the government in his essay *Reflections on Progress, Peaceful Co-Existence and Intellectual Freedom*, it was said in some government circles that he was 'a good bloke, but the Jews led him astray.' The incident with the inkwell, which first brought Veniamin Tsukerman and Lev Al'tshuler together and is referred to in Part 1, was in fact accompanied by an anti-Semitic remark, though this fact is not recorded here.

The role of Beria, too, is passed over lightly. From other sources we learn that this notorious figure was at least a competent organizer, who, in 1952, chose to overlook Lev Al'tshuler's 'incorrect' views when asked to do so by Iulii Khariton. Khariton himself, according to Academician Gol'danskii, made a clear distinction between Beria the organizer and Beria the executioner, although Gol'danskii never heard Khariton pass any adverse judgement on Beria. Khariton is on record as noting that Molotov, who was initially put in charge of the Soviet nuclear programme, was such an ineffective organizer that Kurchatov voiced his dissatisfaction. Beria's son claims that Kapitsa said of Lavrentii Beria: 'It's a pity that such a capable man works for the Bolsheviks.' Samvel Kochariants, for many years Chief Engineer at Arzamas-16, admits that Beria's representatives at Arzamas-16 sometimes impeded the work of the scientists, but says that, by appealing over his head directly to Stalin, such hindrances could be removed. Avraamii Zaveniagin, described by Sakharov as a 'man of great intelligence—and an uncompromising Stalinist', worked at Arzamas-16, nominally as deputy minister to Vannikov, but in fact to oversee security. He held the rank of Lieutenant-General in the NKVD secret police. After Beria was arrested in June 1953 Zaveniagin made a speech at the July Plenum of the Central Committee calling for Beria's removal from the Presidium of the Central Committee. In this speech, which was clearly part of an orchestrated campaign to ensure Beria's downfall, Zaveniagin felt the need to allude to Beria's alleged qualities as a 'worker' (i.e. his organizational qualities as shown at Arzamas-16), and to claim that even these few positive qualities had been 'exaggerated'.

The scientists were materially and financially much better provided for than their fellow-citizens on the 'outside', but the cost was high. Leaving Arzamas-16 was difficult, foreign travel almost impossible, and leading figures such as Khariton and Sakharov were constantly shadowed by Secret Police minders, known as 'uncles' *(diadi)*, or more formally, 'secretaries'. Senior staff had always been responsible for recruitment, beginning with the first seventy recruited by Khariton and Kirill Shchelkin, and any mistakes would be severely punished. An army of informers checked up on staff constantly. Had the project to build an atom bomb failed, there would almost certainly have been executions. Efim Slavskii, at one time Minister of Medium Machine Building, reports that Beria came to the site of the first Soviet atomic explosion, at Semipalatinsk on 29 August 1949, with two lists: one of those to be rewarded in the event of success, the second of those to be arrested in the event of failure. When the project succeeded, rewards were duly showered on the chief participants (see Part 2).

Tsukerman omits to mention, however, that P.M. Zernov was not rewarded, apparently because he had once threatened to strike Beria during a stormy encounter in the Kremlin.

The role of prisoners in building and servicing Arzamas-16 also receives only one mention (Part 2). The Mordovan Republic had always been a prison camp region, and indeed, one such had stood in Sarov itself. In the late 1940s, soon after the opening of Arzamas-16, there was a breakout of some fifty prisoners into the surrounding pine forests, but all the escapers were hunted down and shot. Khariton, too, admits that the subject of prisoners was on the forbidden list: 'People don't talk about it, but they think about it.' It was a phenomenon from which people took 'little joy'. Most of these prisoners were so-called *ukazniki*, that is people who had fallen foul of the numerous decrees *(ukazy)* issued outside the Criminal Code and covering anything from the distilling of illicit liquor to a failure to fulfil the obligatory norm of labour days. For security reasons there were no political dissidents among the prisoners employed at Arzamas-16. Conditions in the 'White Archipelago' were better than in the 'Gulag Archipelago', but on release, prisoners were sent to Magadan, in the Soviet Far East, as far as possible from Arzamas-16. Prisoners working in Arzamas-16 were encouraged to work hard by the piece of doggerel referred to in Part 2:

> Remember this instructive rhyme—
> Work enough, you'll do less time.

Some prisoners ignored this sentiment. Samvel Kochariants tells how his house was broken into and clothes and a clock stolen. Prisoners continued to be used at Arzamas-16 until 1957, when they were replaced by military conscripts.

Another subject which is only touched on by implication is the motivation of the scientists working at Arzamas-16. A cynic might claim that material benefits and superior research facilities were the prime consideration. This was indeed an important factor, as Lev Al'tshuler concedes, but the motivation seems to have been more complex and more principled. Those who have discussed this issue, such as Al'tshuler, Sakharov, and Viktor Adamskii, who assisted in the theoretical department at Arzamas-16 in the 1940s, refer to their experience of the immense suffering of the USSR during the Second World War and the apprehension caused by the American bombing of Hiroshima and Nagasaki. Sakharov's comment, with which Khariton agreed, is typical: 'I expended enormous efforts because I thought I needed to for the sake of global equilibrium. You must understand that this was the only way to avert a third World War'. Gol'danskii makes much the same point in his introduction.

Another motivating factor predates the rise of Nazism. Men like Tsukerman and Al'tshuler, whose father was a Menshevik, were members of an oppressed minority, the Jews, who believed that the 1917 revolution heralded a new world where progress would be guaranteed by science and technology. When the opportunity came to work among the élite scientists at Arzamas-16, where the atmosphere appears to have been conducive to research and, paradoxically, to relatively free thinking, they seized it eagerly. Khariton, incidentally, categorically denies claims made in a recent German publication that the atmosphere at Arzamas-16 was one of alcoholism and cynicism. Russia, unlike the United Kingdom, has never had the notion of the 'Two Cultures' of Science and the Arts and, in Soviet times, scientists often pressed for more cultural, intellectual, social and political freedom. Sakharov is, of course, a prime example, but there have been many other less well-known cases. The poet Andrei Voznesenskii testifies to the intellectual ferment in the nuclear research facility at Dubna in his narrative poem *Oza* (1964) and other writers such as Daniil Granin and Vladimir Dudintsev have chosen the world of Soviet science as their metaphor for Soviet society. In a letter to Izrail Metter dating from 1976 Tsukerman himself insists on the proximity of Art and Science:

> No, your arguments do not hold water. Real science and real art are very, very close to one another. Depending on how they are used, they both can be either moral or immoral.

Veniamin Tsukerman, born 6 April 1913, and Zinaida Azarkh, born 15 May 1917, were not only husband and wife; they were also first cousins. Zinaida's parents were Rakhil (Rachel) née Rubashova and Moisha Azarkh, who russified his name to Matvei. Veniamin's father, Aron Tsukerman, married Rachel's sister Kreina, always known as Katia. Rachel and Kreina were two of the five daughters of Zalman Markovich Rubashov and his wife Zlata. The remaining three daughters, Tsilia (known as Tseita), Khasia, and Rika remained unmarried. The Rubashovs, although Jewish and therefore liable to residential restrictions in Tsarist times, were merchants of the First Guild, a status which enabled the family to live in both Russian capitals, Moscow and St Petersburg. Veniamin and Zinaida met when the relatively impoverished Azarkh branch of the family sent their daughter to Moscow to stay with her wealthier Tsukerman cousin. They married in 1935 and had two children, Irina, born in 1937 and Aleksandr (Sasha), born in 1949. Irina, who is now a Senior Researcher at the Institute of Remedial Teaching (formerly Institute of Defectology) in Moscow, was stricken with tubercular meningitis at the age of nine. The disease was at that time always fatal, but, after herculean efforts by her father and his colleague Izrail Galynker, doses of the drug streptomycin were obtained in the United States, and the child's life was saved, though at the cost of her hearing. Veniamin Tsukerman graduated from the Moscow Evening Institute of machine-building in 1936. Trained

Introduction

as an architect, Zinaida Azarkh became a full-time assistant to her husband in 1941, when he lost his sight. Tsukerman was among the first scientists to be recruited to Arzamas-16. Appointed in early 1946, after having been awarded a Stalin Prize for his work on photographing explosions, he arrived in Arzamas-16 in May 1947 and remained there for the rest of his life. Until his final illness Tsukerman remained very active—he was fond of saying that 'Action is your friend' (Drug—eto deistvie). Over thirty of his pupils obtained doctorates, and nine of them gained post-doctoral qualifications. He was also fond of quoting Saint-Exupéry's comment that 'The greatest luxury is the luxury of human communication'. This informed both his attitude to deaf people and his concern to propagate the arts in Arzamas-16/Sarov. He was a close friend of the painter Varvara Bubnova, whose portrait of him hangs in his grand-daughter's flat. He chaired the Arts Committee of Scientists' House, while his wife chaired the management committee for many years and remains a member to this day. They were responsible for evenings devoted to Ernest Hemingway, Marina Tsvetaeva, Garcia Lorca, Anna Akhmatova, the art and poetry of Japan and the plays of the archetypal playwright of the *perestroika* era Mikhail Shatrov. Apart from literature, music played a great part in Tsukerman's life from his earliest years. Drawing an analogy between it and his chosen profession, he wrote to a friend: 'Don't abandon music. Its role in, and influence on, the life of man is much more long-lasting and indivisible than say, the stability of the proton.' Veniamin Tsukerman died on 25 February 1993 and is buried in Sarov. At his death he was the holder of four State Prizes, a Lenin Prize and the titles 'Hero of Socialist Labour' and 'Honoured Inventor of the RSFSR'.

Tsukerman's interests extended well beyond nuclear physics. His research on radioisotopes, for example, found practical application in the Soviet space programme. He also had a strong interest in medical research, prompted, perhaps by the fate of both his children, and his own blindness which was said to be hereditary. In 1962 he was instrumental in establishing a small radiobiology group at Arzamas-16. Later, as this book recounts, he devised and built the oxygen chamber which saved the life of his colleague's wife, Liudmila Golubeva. In 1984 they co-authored the book *I Cannot Hear (Chelovek ne slyshit)*, though, for security reasons, authorship was ascribed to 'Z. and V. Krainin'. The book, published in an edition of 200,000 and reissued in 1987, gave the first popular account to appear in the Soviet Union of the problems associated with deafness. Tsukerman had already been instrumental in establishing a research laboratory at the Institute of Defectology which developed equipment for deaf and deaf-blind people. In 1966 his campaign to have television programmes subtitled or interpreted into sign language bore fruit, and he also pressed the authorities, through newspaper articles, to improve the lamentable provision of hearing aids.

Both the authors and the editor are conscious that there are passages in this book devoted to specialized technical information. As the authors say at the beginning of Part 2, the reader who is not technically inclined should not feel discouraged. There is much in this book to interest everybody.

The Library of Congress system of transliteration has been used, with some modifications in the interests of clarity, such as the disregarding of final soft signs and the simplifying of names such as Maria and Lidia (rather than Mariia and Lidiia). Most of the photographs appeared in the Russian edition and were taken from the archives of Zinaida Azarkh and Iulii Khariton. I am most grateful to David Gillespie, to Rosalind Marsh, who read the original translation and made many helpful suggestions, to James Muckle, who encouraged the project, to Wendy Oldfield, who keyed in and printed out numerous drafts of this book, to my other former colleagues in the Russian Department at Exeter University, who, despite being under considerable pressure from a largely unsympathetic administration, allowed me to make several research visits to Moscow, to Richard Rhodes and to Anna Komarova, her mother Irina Tsukerman, her grandmother Zinaida Azarkh, and Aleksei Iur'evich Semenov, grandson of both Iulii Khariton and Nikolai Semenov, who collectively and individually answered my numerous queries with great patience and accuracy. Any remaining errors or omissions are entirely my responsibility.

Michael Pursglove   Bath, October 1998

Behind the Iron Shutters of Secrecy

by *Academician V.I. Gol'danskii*

THE FATE of science is dramatic. We realize this most often when we contemplate the fate of a particular scientist or a particular invention. However, the history of science contains pages which are particularly dramatic, when a great number of people are sucked into the orbit of an idea about to be realized, when the fate of a state and, perhaps the fate of the world depend on the potential and the level of many spheres of basic science, on the talent, preparedness and self-sacrifice of both researchers and those charged with practicalities. One such page is open before us in the documentary narrative of Veniamin Aronovich Tsukerman and Zinaida Matveevna Azarkh.

Up to now only the basic facts have been known. On 6 August 1945, three months after the end of the war with Nazi Germany, an American atom bomb fell on the Japanese city of Hiroshima and on 9 August another fell on Nagasaki. By no means everyone realized that we were in a completely new situation. The balance of power had shifted; the consequences of this shift could only be far-reaching. With its nuclear blasts the United States had demonstrated to the entire world that it alone possessed super-weapons and was ready to use them if, from its standpoint, the need arose.

In order to eliminate this one-sided threat, in order to remove the danger of nuclear blackmail, the USSR had only one way out—to create an equally powerful weapon itself.

Today we know what happened: only four years later, on 29 August 1949, the Soviet atom bomb was successfully tested. Moreover, a further four years after that, the USSR was in possession of an even more powerful weapon—the thermonuclear bomb. The American nuclear monopoly was at an end. And it can be asserted with confidence that at this stage of history 'nuclear restraint' played a positive role in averting a third world war.

This was an outstanding event in the history of science and the history of society. But how was this achieved in an amazingly compressed timescale in a country devastated and drained of its life blood? It is precisely this—how the impossible was achieved—that until recently none of us knew. All information about it was top-secret and the leading players anonymous. Soon we learned the name of the Head of the Soviet nuclear project: Academician Igor Vasil'evich Kurchatov. Much later the name of the builder of the first Soviet atom bomb—Academician Iulii Borisovich Khariton—became known, as did the major role played by Academicians I. E. Tamm, A.D. Sakharov, Ia. B Zel'dovich and others.

How was all this achieved and at what cost? The documentary narrative of two of those who participated in the events—V.A. Tsukerman and his wife and assistant Z.M. Azarkh—is one of the first glimpses behind the iron shutters of secrecy.

Both in the authors' introduction and in the concluding lines of the narrative we can hear the questions which, above all others, trouble the authors: will these memoirs be of interest to young readers? Will our problems and anxieties excite them? I make so bold as to reply—they will. They will, because we are dealing with truly remarkable people and because what they did was of truly historical significance; they will, because the book has not been written by peripheral observers but by active, creative participants of whose feelings and thoughts the reader is conscious throughout.

Many of the people who figure in this narrative, as well as many of the events described, evoke memories in me. I am therefore able to check the images on these pages against my own impressions and to vouch for their accuracy. Thus it became doubly interesting, pleasant—and also nostalgically melancholic—for me to read these pages.

I think that anyone who played any role at all in the events described will feel this sense of nostalgia. I have, however, to record with sorrow, that many of them died without living to see this open acknowledgement of their work.

The earliest events in the narrative are linked with the war years when, after the battles of Stalingrad and Kursk, the tide had turned in 1943 and the country was able, with great difficulty, to set about the task of acquiring atomic energy.

At this point it must be said straightaway that this work did not appear out of a clear blue sky. The foundations had been laid by the outstanding achievements of Soviet science in the pre-war years. They ensured the success of the whole undertaking. Unforgivably, this fact is often forgotten by the officials and bureaucrats of today. It is painful to see how Russian fundamental science has abandoned its position in the world because the state economic apparatus does not want to understand, or is simply incapable of understanding, that the task of science is to investigate the laws of Nature and society, to acquire new knowledge and not merely to apply science to economic practice. This problem is vitally important for the state, and it is no accident that I have paused here, bearing in mind the history of the creation of the atomic bomb. It was precisely such fundamental work on the theory of explosion and detonation and on the division of isotopes, profound, variegated chemical research, which allowed us in 1943 to develop research in the sphere of atomic physics and atomic energy. Without this foundation, without the ground-breaking work of N.N. Semenov, I.V. Kurchatov, Ia.B. Zel'dovich, G.N. Flerov, K.A. Petrzhak and V.G. Khlopin—no efforts, however titanic, would have led to such rapid success.

Of course Tsukerman and Azarkh have not touched on all the problems and have passed over the 'subdivisions' of the army which tackled the problem. Beyond their narrative remains the solving of many physical, chemical and engineering problems connected with nuclear charges. However that would have been simply impossible in a narrative of this nature, when the authors are sharing with the reader what they themselves lived through. Perhaps the most important and attractive feature of the book is the fact that the authors, basing themselves on their own experience, have recreated the atmosphere which was characteristic for all the groups which took part in the solving of the atomic problem. In a shattered country, where there was insufficient electricity even for lighting, where original pieces of equipment literally had to be 'cobbled together' from almost nothing, oblivious of their tiredness, people created things from whatever was to hand, such as an old transformer lying about in the street or a mirror cadged off a barber. In their relations with one another people not only shared their thoughts but gave each other firm, reliable support, irrespective of their decorations and rank.

There were no bureaucratic hassles. Every invention not only avoided excruciating trials but was introduced there and then and evoked general rejoicing. It is precisely for this reason that, in telling the story of experiments on explosions and detonations, the authors give equal due to the Academician, the glassblower, and the people who ensured that life ran smoothly.

Efforts such as these are not easily forgotten. In surveying those tense months and years the authors cannot help but recall not only the final victory, which was regarded in the west as tantamount to a miracle, not only Beria's repressive régime which hung menacingly over everybody, but also the atmosphere of kindness and mutual aid which enabled the miracle to happen. Unfortunately this atmosphere is rarely, if ever encountered in the bureaucratized scientific system of today. Perhaps it is time that we remembered that, in the absence of such an atmosphere, the shoots of new ideas simply wither.

The years passed and the day came when one of the creators of the hydrogen bomb, Andrei Dmitrievich Sakharov, a man who, in his own words 'had had no doubts about the vital importance of Soviet nuclear weapons both for our own country and for maintaining the world in a state of equilibrium', became one of the first to come out in favour of the reduction and gradual elimination of nuclear weapons. What was this—a contradiction, a betrayal of self, as one sometimes hears? No, it was, of course, part of the debate about the development of science and civilization and showed the ability of one genuine scholar to understand and acknowledge it.

Such is the logic of history. At that moment in time atomic weapons were simply vital and their creation represented a massive achievement. Now, in whoever's hands they may be, they are dangerous and it is vital that we eliminate them. The words *Sic transit gloria mundi* might be applied, except that these weapons could never be called glorious. On the other hand the adjective could be applied to those timeless pages of our scientific history, which until recently were kept under wraps, and to those people whose names are forever imprinted on those pages.

Today the movement against not only nuclear weapons, but also against nuclear energy in general, is gaining strength throughout the world. History, however, teaches us that in critical global situations we must not lose our heads and give in to the herd instinct.

Of course the accumulated mass of nuclear weapons can, by the mere fact of their existence, pose a threat in conditions of mutual distrust. However the new thinking, the new relationships between the USA, the USSR and other countries which have grown up in recent years, the abandonment of traditional hostilities, the general striving towards mutual understanding open new possibilities for turning explosive into fuel, for turning swords into ploughshares.

That is why it is vitally important to strengthen the thread of trust which stretches between peoples across oceans and continents. The whole future of our civilization depends on our using the potential of nuclear energy rationally and in harmony, without threatening humanity with an apocalyptic nuclear war or the agonizing slow death by radiation caused by hundreds of Chernobyls. Therein lies the answer to the question 'to be or not to be' in relation to our planet and our common home.[1]

---

[1] Both 'our common home' and 'the new thinking' in the previous paragraph were catchphrases of the Gorbachev years.

Foreword

THIS book tells the story of its authors' lives and their decades of participation in the creation of Soviet atomic weaponry. It also tells of the remarkable people who worked alongside them. The book has two authors, although it is narrated for the most part in the first person singular. This form turned out to be more convenient for setting out the biographical and other material which they had at their disposal.

The advent of the nuclear age opened up ways for humanity to exploit gigantic sources of nuclear and thermonuclear energy, and at the same time created the threat of nuclear catastrophe. The entire world experienced the reality of that terrible threat with the destruction of Hiroshima and its people by a single American bomb.

The doors of the atomic era began to open in early 1939, when the German physicists Otto Hahn and Fritz Strassmann, together with Lise Meitner and Otto Frisch, discovered the breakdown of uranium atoms into two fragments of similar mass and into neutrons which relayed the fission reaction to other atoms. Very soon, during the period 1939-40, the Soviet scientists Iakov Zel'dovich and Iulii Khariton—both major authorities on detonations and explosions—established the theory of the chain-reaction decay of uranium in a series of articles. This process, which leads to an enormous release of energy, could, as the authors indicated, be used for a wide variety of purposes. Similar work was being done in a number of other countries. In July 1940 Igor Kurchatov and Khariton became members of the Uranium Commission organized by the Presidium of the Academy of Sciences. In August of the same year, together with Lev Rusinov and Georgii Flerov, they sent a letter to the Academy of Sciences entitled 'On the use of energy from uranium in chain reactions'. Their work was interrupted by the outbreak of World War II. In 1943 the situation on all fronts had improved to such an extent that we were able to resume our research work in atomic physics and thermodynamics.

The tasks facing us were enormous and, in the conditions prevailing at that time, almost insuperable. The Russian nuclear project required the creation of a whole new branch of our mining industry in order to extract the necessary raw materials. In order to obtain new elements, immensely complicated reactor installations had to be created, along with the technology needed to process these elements into weapons-grade materials. Scientific leadership of the entire project was entrusted to Kurchatov, while the enormous responsibility for building the atomic bombs was given to Khariton.

Information passed to the Soviet Union by the prominent German physicist Klaus Fuchs, who had taken part in the creation of the American atomic bomb, first successfully tested in 1945, also facilitated the successful outcome of the Soviet nuclear project. However, according to the British scientist Rudolf Peierls, it is unlikely that this intelligence data greatly affected the course of events, and scientists estimate that it brought forward the development of the Soviet nuclear bomb by no more than a year. Credit for the creation of our hydrogen bomb belongs entirely to Soviet scientists, foremost among them Andrei Sakharov. We had no intelligence information on this from American sources.

Fuchs himself characterizes the significance of the information he passed on thus:

> ...It is difficult for me to judge just how valuable and necessary my information was for the Soviet nuclear programme. After all, at the end of the 1940s the Soviet Union was conducting its own nuclear weapons development programme on a broad front, engaging the whole of its industrial and scientific resource base. Despite the carefully preserved nuclear monopoly of the United States and a total blockade of scientific information from the West, Soviet nuclear scientists achieved a great deal by dint of their own efforts. The information I passed on to Soviet representatives was valuable, in my view, above all insofar as it helped Soviet scientists avoid unfruitful lines of research and concentrate their efforts in the most important areas.

We had, of course, to check Fuchs' information and make sure that it was not disinformation. Before us lay a huge experimental and theoretical task.

Foreword

Outside Moscow, in the town of Sarov, a large industrial and experimental complex was built. Here delicate physics experiments could be conducted, complex mathematical calculations could be done, multi-ton explosive charges could be detonated in order to conduct precision measurements of the fleeting processes involved, and the properties of materials could be studied under pressures of millions of atmospheres. Highly qualified theoreticians, experimenters, engineers and constructors were invited to the Sarov complex to do this work. Pavel Mikhailovich Zernov was appointed director of the new complex and Iulii Khariton became chief engineer.

In conditions of the utmost secrecy thorough tests were soon performed on the first variant of the atomic bomb; next came the development of original and much more effective types of nuclear weapons, and, after 1953, of thermonuclear weapons. In order to ensure maximum reliability, a type of bomb similar to the 1945 American bomb was used for the first test in August 1949.

The Sarov complex saw the development of new research, including the creation of an entirely new branch of physics, the physics of high-energy densities. In a country half destroyed by war this was nothing less than a miracle. Significant contributions to the solution of these problems were made by Iakov Zel'dovich, Andrei Sakharov, Igor Evgenevich Tamm, Evgenii Ivanovich Zababakhin, David Al'bertovich Frank-Kamenetskii, and many experimenters at Arzamas-16: Kirill Ivanovich Shchelkin, myself, Lev Vladimirovich Al'tshuler, Viktor Aleksandrovich Davidenko, Mikhail Vasil'evich Dmitriev and others.

In this book we have decided not to touch on the physical, chemical and engineering aspects of the functioning of nuclear charges. On the other hand, certain problems of general physics, such as the behaviour of substances under superhigh pressures, the measurement of pressures on the detonation and shock-wave fronts, and so on, take up considerable space. There are many problems connected with the development of nuclear weapons which the book does not claim to embrace, but it does tell the story of experiments and research in which the authors participated directly. Many scientists and researchers who made significant contributions to the development of nuclear armaments either do not appear at all or are only mentioned in passing. In these cases the authors considered themselves too little acquainted with the people concerned to have the right to act as their biographers.

At the present time, when the entire civilized world is talking about the banning and elimination of nuclear weapons, the efforts, the enthusiasm and the energy which were put into the creation of Soviet nuclear armaments may seem out of place. However, almost half a century ago we were facing a different situation. It was imperative to end the US monopoly on new weaponry as soon as possible. Doing so served the interests of peace not only for the USSR but for the entire world.

A great many friends and colleagues provided invaluable assistance in the writing of this book. It is the authors' pleasant duty to express cordial thanks to E. and G. V. Aleksandrovich, L.V. Al'tshuler, E.I. Arsen'ev, E.M. Barskaia, S.M. Bakhrakh, A.M. Voinov, G.B. Voinova, V.N. Beliaev, M.F. Kovalev, V.A. Nazarov, V.I. Nemyshev, R.Z. Liudaev, I.Sh. Model, N.G. Pavlovskaia, M.V. Sinitsyn, V.Ia. Frenkel, L.N. Khudiakov, N.D. Iur'eva.

PROLOGUE

> Each man writes the things he hears.
> Each man hears the way he breathes.
> The way he breathes is how he writes,
> Not trying to please another's taste.
> That's what Nature has demanded.
> Why she has is not our business,
> Her purpose is not ours to judge.
> *Bulat Okudzhava*[1]

Tap-tap. Tap-tap-tap. Tap-tap. The sound of a woodpecker. It is sitting on a tall pine tree and tapping the trunk with its beak. Very like the sounds of the old Morse code telegraph. True, the woodpecker only does dots. It transmits its signals slowly, like a trainee radio operator. There, the tapping has stopped. The bird has flown to the next tree. The tapping resumes, but much more faintly now. The pine he's chosen now for his transmissions is further away from the window. A typewriter begins to tap, as if answering the sound of the woodpecker. He has fallen silent. Perhaps he was taken by surprise. However, suddenly his signals get louder. He's set up operations right under the roof of the wooden house where we've been living for more than a quarter of a century. Besides the tapping exchanges with the woodpecker you can hear a nightingale singing somewhere very nearby. But this happens only from mid-May to early June. The woodpecker stays on transmission duty the whole year round.

To the right of the typewriter is a pile of white paper. As I find the words I seek, a pile of typed pages grows to the left of the typewriter. Thus this book takes shape. In the course of the extraordinary life which I and my friends have lived a great many

---

[1] The final verse of 'I am writing a Historical Novel' ('Ia pishu istoricheskii roman', 1975) by Bulat Okudzhava.

xxxiii

things have happened which I should have set down a long time ago.

When you start a new manuscript, you always feel a certain apprehension. Will it be interesting? Will people find it useful? The events and facts which make up its content may be of little interest to 'the unknown tribe of youth'.[2] They haven't been through what we've been through. 'Titans of the human spirit'—that's what you could call many of the people with whom I have had the good fortune to work for many decades.

I have been wanting to start writing for a long time now. I already have some passages roughed out. Individual phrases and paragraphs torment me when I can't sleep. I have to commit them to paper while there's still a chance before they dissolve in the stream of everyday concerns.

It is difficult to assign this book to a particular genre. It will contain a great deal about the new science which is both daughter and heiress of a remarkable century which will leave in its wake a great many more questions than answers.

The woodpecker has stopped tapping. On the other hand, the typewriter is picking up speed. And the eternal question—'Will it work?'—looms over my desk. I'd like to think that it will.

---

[2] A quotation from the poem 'Again I visited' ('Vnov' ia posetil', 1835) by A.S. Pushkin.

Part 1

VITEBSK — MOSCOW — KAZAN — MOSCOW

From this autobiographical chapter the reader will learn how we lived in the 1920s and 1930s, how we came to be involved in the greatest adventure of our lives, and how we managed to work in Kazan during the difficult war years.

MY VITEBSK (1913-28)

> Have you really been to Vitebsk?
> Do you mean to say you have
> actually been to Vitebsk?
>
> Iu. Trifonov, 'A Visit to Marc Chagall'[1]

IF Marc Chagall had asked this question of me, I would have answered him unequivocally. Yes, I was born in Vitebsk and spent the first fifteen years of my life there. There were even rumours that Chagall was a distant relative of ours. But all the members of our family who knew our genealogy were either killed during the war or else died of natural causes.

Vitebsk was founded in 1109. In the 1920s its population was slightly over one hundred thousand. At that time Minsk was about the same size. There was even a celebrated debate at the time over which of the two cities should be made the capital of the Belorussian Republic. Minsk was the choice, which caused great offence to the people of Vitebsk. The Second World War changed a good deal. When we returned to Vitebsk in 1958 I was astonished at the extent of the destruction which had occurred

---

[1]This short story (Russian: 'Poseshchenie Marka Shagala') was first published in the journal *Novyi mir*, no. 7, 1981, 84-87.

there. Nothing remained of my old school, a massive three-storey building on the banks of the Western Dvina. Frunze Street, including house no. 16 where we lived from 1917 to 1928, had been made into a boulevard. By some quirk of fate the two-storey brick building where I was born had survived. It stands to this day on Dimitrov street, opposite the entrance to the Tikhantovskii Garden—that was the pre-revolutionary name of the park where the summer theatre used to be.

My memory has preserved my first childhood impressions: a regiment of soldiers marching down the street, led by a band and singing a popular song of the time:

> Weapons in the sunlight gleaming,
> And sounds of dashing martial brass,
> In clouds of dust, moustaches streaming,
> A regiment of hussars pass.

I learned to read at the age of six. In 1919 the schools were not operating and I used shop signs to teach myself. The first books which I read on my own were *Uncle Tom's Cabin* and *David Copperfield*. These books became a lifelong source of inspiration for me. They made me a confirmed internationalist and instilled in me the importance of doing good to others.

Music... In the room where these lines are being typed there is an old cherry-coloured piano, a Türmer. It is a few years older than I am. The front face of the keyboard lid is carved with bunches of cherries. This piano has followed me from city to city for over seven decades. Even so I never learned to read music. I was prevented by my early interest in technology, which left me with practically no spare time. However, I learned to play by ear tunes from operettas, fashionable waltzes, and mazurkas.

My mother was a gifted musician. She knew a great many Russian, Ukrainian and Belorussian songs, most of which found their way into my 'repertoire'.

There was another thing which prevented me from becoming a musician. The boy next door, Mark Fradkin, had perfect pitch. He and his mother lived on even more modest means than our family. They did not own any musical instruments and the future composer and songwriter often used to come round to our place to play the piano. When he went to see an operetta, Fradkin would memorize all the parts and then reproduce them on the piano with complete left-hand accompaniment. My own ear was only good enough to enable me to grasp the main melody and play it back in the right hand. Nevertheless the cherry-coloured piano has helped me to get through the tragic situations of which I've had more than my share in the course of a long life. Even today, when I'm over seventy, this piano, this old and faithful friend, helps me to carry on.

No-one knows how or why it is that music, even music played imperfectly by ear, exerts such an influence upon a person. I discovered this influence while I was still a child. Then I had an idea: I would assign my friends their own particular tune. My friend Leonid[2] was assigned his favourite Scherzo No. 1 in G flat minor by Chopin. Academician Evgenii Zababakhin was associated with Beethoven's Moonlight Sonata and with the wartime song 'Oh! The Roads!' Whenever he came to visit us he would usually sit down at the keyboard and play one or other of these pieces. 'Music ought to strike fire from the human breast.' I often think about these words of Beethoven.

I developed a passionate interest in technology from an early date. By the bridge over the Vit'ba River (from which the town took its name) there was a small power station which supplied the tram system and other municipal facilities. My route to school took me past the power station. I could stand for hours outside the brightly lit windows of the power station and watch the duty engineer at work. The station was equipped with a steam engine, the revolution rate of which was maintained by a Watt centrifugal regulator. I knew how it worked from Zinger's *Physics*, a book I had read with the same delight as *Uncle Tom's Cabin* and *David Copperfield*. It was always a great moment when the engineer allowed us boys to oil the bearings of the steam engine with a big oil can.

In the summer of 1922, when I turned nine, and my younger brother was only four, our father died. My mother[3] was left alone to bring up two small boys. Life was hard. We understood

---

[2] A reference to Izrail Solomonovich Galynker, known as Leonid, a metallurgist and childhood friend of V. A. Tsukerman. He obtained from America the drugs necessary to save the life of Tsukerman's daughter, Irina, and, as a result, was sentenced to death for 'an attempt on the life of Comrade Stalin'. The sentence was commuted to twenty-five years' imprisonment and he was 'rehabilitated' in 1955. Among the 'evidence' against him were the phone calls made to America and Sweden in the race against time to save Irina Tsukerman.

[3] Tsukerman's parents were Aron Tsukerman and Kreina (known as Katia) Zalmanovna née Rubashova. His brother was Samarii Aronovich, known as Shura.

3

that and I did not always pluck up courage to ask Mother for twenty kopecks for sheets of paper to use in cut-and-paste projects. Then I took my first independent steps in electrical engineering. There was no electricity in our flat. It took me almost six months to save up enough money for wiring, cleats, switches and lamp-holders. How to connect the switch—in parallel or in series—was the first engineering problem I ever had to solve. It arose immediately, as soon as I started the work. I thought it over for more than a day, then finally I understood: the switch had to be connected in series with the lamp. When at last there was a light bulb burning in our flat, the joy of my mother and my brother knew no bounds.

I began inventing things early in my life, when I was six or seven. I would build little cottages and houses out of dominoes. At that age the feeling that I 'should do things differently from other people' was already well entrenched in me. The year my father died I was given a construction set—a collection of plastic pieces with special apertures in them, with nuts and bolts to match, out of which you could make all sorts of machines and mechanisms. This construction set was to be of enormous significance in my life. For some five years it was my constant companion. I built the models according to the instruction book, then I began to build my own inventions. One of them nearly cost my younger brother an eye. With the help of the construction set I built a working model of a centrifuge cannon. It was very difficult to aim it. When I was test firing it one day, a centrifuge shell hit my brother in the eye. It was only by lucky chance that his sight was not damaged.

I had another childhood passion—aviation. Not far from our house was a ravine and beyond it they had built a small aerodrome. Boys from up and down the whole street would gather there to watch the planes. At that time aeroplanes were unlikely-looking contraptions made of plywood and fabric. As a rule, they were biplanes. They flew slowly and their range was short. But what a thrill it was to watch one of them hurtle a short way down the runway and then lift off into the air! The boys would spend the whole day there, from morning till late at night. As a class assignment for my physics teacher, I gave a paper on the principles of heavier-than-air aviation. I illustrated my talk by launching a working model. When he met my mother, my physics teacher told her: 'Your boy clearly has a talent for science. We must develop his knowledge of mathematics and physics.'

From 1925 onwards I took an active interest in radio technology. I built my first crystal receiver set in 1926. In the same year I became a member of the FRS (Friends of Radio Society). In the course of many years' work in various branches of physics and technology, it often happened that some device of mine performed successfully. But when I first touched the needle of my receiver against various points on the crystal and heard those precious words: 'This is Moscow. You are tuned to Comintern Radio', the feeling was indescribable.

'Vitebsk, farewell!' I could have repeated these words of Marc Chagall when, in 1928, after completing seven years of schooling, I left for Moscow to continue my studies

FRIENDS OF THE 1930S

I MET Lev Al'tshuler at the end of the 1920s. At that time there was no secondary education beyond the age of fourteen. After seven years' schooling boys and girls could continue their studies for another two years on what were called 'special courses'.

In Moscow I enrolled on such special courses in technical drawing and design, with an emphasis on construction. After two years' study, including practical experience, students enrolled on such courses received both their school-leaver's certificate as well as their diplomas as junior foremen or designers.

I first noticed Lev in September 1928, during the first days of my studies. During break one day one of the students got into a heated argument with him over something.[4] Within a few seconds an inkwell sailed across the classroom and smashed against the opposite wall, leaving a large ink-stain on the wall. I liked the speed of his reactions. Someone like that will always be able to stand up for himself. Our friendship began with that inkwell.

A number of episodes from those years have left very distinct impressions on me. In 1929 we were working on the construction of the 'House on the Embankment', a building which gave its

---

[4]Tsukerman omits to say that the incident arose from an anti-semitic remark made by a fellow, non-Jewish, pupil.

5

name to one of Iurii Trifonov's best works.[5] The building is still there in Moscow today and contains the well-known Udarnik cinema. In the early thirties there was no building technology available in Russia and students doing their practical training were used mainly for moving bricks, with the help of simple contraptions known as 'nanny-goats'. When I was doing my practical training, on a six-storey building near the Semenovskii Gates, things they called slewing cranes appeared. These were fairly primitive machines which allowed you to hoist bricks, cement, mortar and other materials vertically, and then shift them horizontally, too, within a range of about ten metres. Once the crane operator suggested that I climb up on to the uppermost girder in order to lubricate the top bearing on the crane.

'When you're on the girder, don't look down or you'll be in trouble,' he warned.

I couldn't resist the temptation, however, and when I got to the middle of the girder, I looked down. He was right. My head started to spin, but I got away with it, did what I had to, and got down safely.

I spent almost my whole practical training working as a crane driver. Soon after I got my certificates, significant changes took place in my life. These stemmed from my other passion—radio. I did not have enough money for books so I often went to the Lenin Library to read the journal *Radio Front* or other publications on the same subject. When I was engaged in this one day, a grey-eyed, fair-haired man of about thirty came up to me, gestured towards the books strewn round on the table, and said: 'I've been watching you for a long time. You work in radio, do you?' 'No, radio's just a hobby of mine.' 'I'm looking for an assistant to work in my radiographic laboratory. Judging by what you read, I reckon you'd be right for the job.'

That is how I met Evgenii Fedorovich Bakhmetev. He was then a young, dynamic professor, a graduate of the Zhukovskii Air Force Academy and a pupil of Professor Geveling. Bakhmetev was at the time in charge of the Aviation Materials Testing

---

[5]Trifonov's novel *The House on the Embankment (Dom na naberezhnoi*, 1976) is a modern classic, depicting the pernicious effects of Stalinism on Russian society. In this context a 'house' (Russian: *dom*) is a large block. The house referred to here contains, in addition to the cinema, a supermarket and a bank, and other commercial premises. In Stalinist times members of the governmental élite lived here.

Division of the Central Aero-Hydrodynamics Institute, as well as organizer of a radiographic teaching laboratory in Moscow's first mechanical engineering evening institute.

Bakhmetev's biography could occupy a whole separate book. Like many academics, following the murder of Kirov in December 1934, he was arrested an exiled from Moscow. For many years he was forbidden to live in the major cities of the Soviet Union. After settling in Kostroma he began to work in one of the departments of a textiles institute. With the consent of the director of that institute, we made some special radiographic apparatus for Bakhmetev's research into fine-grained fibrous structures. But the Second World War forced radical changes to all these plans. Bakhmetev was arrested again and sent to Kazakhstan. Our family was evacuated to Kazan and we lost touch with him. Once we were back in Moscow we tried in vain to reestablish contact with him. At last we managed to find out that in Kazakhstan Bakhmetev had fallen on extremely hard times and in 1944 had died of a heart attack complicated by undernourishment.

Bakhmetev offered me a training stint in his laboratory on the understanding that, after a year, the Evening Institute would have its own building which would house all the laboratories. I agreed to this and in 1930 I became an assistant in the radiographic laboratory.

In those far-off years there was still no industrial radiographic equipment in Russia. Production of medical X-ray equipment had only begun in 1928. Nor were there any trained radiologists. I worked for about a year in the Aviation Materials Testing Division, which was soon renamed the All-Union Aviation Materials Institute. In 1931 I was given a large room in the basement of one of the buildings on Blagoveshchenskii Lane, and I set about organizing a radiographic teaching laboratory. In the autumn of 1931 the laboratory took in its first students. Until 1935 this was the only teaching laboratory in Moscow for research into industrial radiography. Students at the Moscow Bauman Higher Technical School and the Institute of Non-Ferrous Metals and Gold, whose curricula included courses in fault detection and radiographic analysis, passed through our laboratory.

At that time, besides Bakhmetev, there were four other people working in the laboratory: a lab assistant Nikolai Georgievich Sevast'ianov, an aeronautical engineer Aleksandr Fedorovich

Sinitsyn and two other lab assistants. We had to do practically everything ourselves, not only soldering and welding, but glassblowing and making vacuum equipment as well. Half a century later I recall all these teachers with gratitude.

First Inventions

THE period 1930-31 was also the time of my first inventions in radiographic technology. At this time, too, the director of the radiography division of the Dutch firm Philips proposed and created a rotating anode for increasing the power of X-ray tubes. The idea occurred to me of rotating not the anode but the whole tube, at the same time deflecting the electron cluster to the periphery of the anode with the help of a powerful electron magnet. I came out with this proposal at one of Bakhmetev's seminars. Bakhmetev was interested and advised me to go to the Svetlana factory in Leningrad, which was, at that time, the only one in the country producing X-ray tubes. The Head of the Radiography Division of Svetlana, F.N. Karadzha, approved my device, which increased the power of an X-ray tube by almost a whole order of magnitude, and advised me to apply for a patent on it. I followed his advice, but a few months later my application was rejected. It turned out that as early as 1896—a year after the discovery of X-rays—Thomas Edison had invented a similar device. I wasn't particularly downcast; however you looked at it, I was competing with the great Edison himself.

Lev Al'tshuler also found himself in the radiographic laboratory more or less by chance. In 1930, after completing his 'special courses', he was assigned to the Volga region to work on the construction of animal breeding facilities. After a two-year stint in one of the collective farms, he returned to Moscow. I ran into him unexpectedly on Tverskaia Street. I hauled him into the radiography laboratory and showed him how levels of high voltage based on the gap between two spherical dischargers were measured and how a luminescent screen glowed under X-ray bombardment. I introduced him to Bakhmetev and soon Al'tshuler, too, became an assistant in our radiography laboratory. I can distinctly remember his first encounter with Bakhmetev:

'Tell me, Evgenii Fedorovich, is there any chance of making a major discovery in your lab?'

'Probably, provided you work hard enough,' replied Bakhmetev.

Soon afterwards another young talent joined the staff. In the autumn of 1932 a close friend of Bakhmetev's, Nina Konstantinovna Kozhina introduced into our lab a sixteen-year-old who had just completed seven years of schooling and did not know where to apply for further education. This was Vitalii Lazarevich Ginzburg, who would later become a famous physicist, Academician and director of the Theoretical Division of the Physics Institute of the Academy of Sciences (FIAN). There were no vacancies and for the first few months Vitalii worked for no pay. After graduating from Moscow State University he became a postgraduate in the physics department of the university. His supervisor there was Academician Igor Evgen'evich Tamm.

At that time Al'tshuler, Ginzburg and I formed the core of the radiography lab. Since the Evening Institute did not have premises of its own for a long time, the laboratory moved several times. After Blagoveshchenskii Lane we moved to the Artem Workers' Faculty[6] on Bol'shaia Ordynka, then to Vuzovskii Lane on the Boulevard Ring Road and finally, in 1938, to Shabolovka. By then our laboratory was actively involved in techniques and apparatus for radiographic structural analysis and industrial radiographic defectoscopy.

During its first years the Evening Institute was headed by a man named Bondarev, a worker who had been specially promoted from the rank and file. At his insistence I enrolled in the Institute's second-year programme in the cold metalworking department. I completed the programme in 1936. The theme of my dissertation was instrumentation for determining the quality of cold-worked metal finishes. I invented a modification for metallurgical microscopes that would give a profile view of finished surfaces at 1000x magnification. I remember very well what the Chairman of the Examiners, Professor Sergei Sergeevich Chetverikov, said in his summing up at my viva voce: 'If this candidate had passed the requisite courses for a doctorate, I would have suggested that he offer this thesis project as a doctoral dissertation.'

---

[6]Russian: *rabochii fakul'tet* or *rabfak*. A preparatory department in higher educational institutions, which prepared workers who lacked the normal entry requirements for admission to higher educational courses.

In the same year, 1936, Lev Al'tshuler was completing his studies in the physics department of Moscow University. It had taken him only three years to complete his degree instead of the usual five.

During the First Five-Year Plan, an era well known as the 'era of industrialization', new plants were being built with their own laboratories. In addition to our laboratory, new labs sprang up in a number of institutions, among them the Zhukovskii Air Force Academy, the Moscow Institute of Non-Ferrous Metals, and the Bauman Higher Technical School. Depending on the power and importance of these institutions, the basic equipment for their laboratories was either acquired abroad or else assembled from components which had had some other designation, for instance from diagnostic or therapeutic medical X-ray equipment. There were few radiographic technicians and I received many offers to assemble or install equipment for radiographic structural analysis and industrial defectoscopy.

In 1935 we received just such a proposal from the management of the Sergo Ordzhonikidze Machine Tool Plant. They had just bought some radiographic equipment from the German firm Siemens. When they tested one of their new installations, all its elements had performed normally, but the X-ray intensity was for some reason almost a whole order of magnitude less than the value given in the specifications. We spent two solid days solving this puzzle. It turned out to be a matter of reversed polarity on transformers, and we were able to sort things out ourselves.

American and German publications in the field of impulse radiography, which appeared between 1938 and 1941, led to the development of this line of research in our laboratory as well. What was distinctive about our work was the use of a standard kenotron as a source of X-ray flashes of only microseconds' duration. This rectifier tube had a large tungsten anode and a powerful cathode which could tolerate significant short-term overheats.

Lev Al'tshuler was soon called up to serve in the army and I was left to work in the lab with Aleksandr Ivanovich Avdeenko, who had joined us in 1937. Before that he worked in various radio-technological institutions, including the famous radio

Vitebsk - Moscow - Kazan - Moscow

laboratory set up by M.A. Bonch-Bruevich in Nizhnii Novgorod.[7] He took part in the development of the first Soviet oscillating and receiving radio tubes. He was an inventive man who had completely mastered the most varied of skills— glassblowing, soldering, electric welding, and photography, about which he was particularly enthusiastic. He was an interesting man, too, on account of his gentle Ukrainian humour. When his twin daughters were born, he informed us cheerfully: 'How nice—two at once. One for the developer and one for the fixer.' His reaction to my proposition on the need to help people and to treat them kindly: 'What sort of a boss will you be, Veniamin Aronovich, if you don't do anything nasty to anyone?'

During this period we were working on X-raying a steel core in free fall, using an overheated kenotron. We perfected this technique to such an extent that in March 1941, three months before the outbreak of war, we were able to take the first X-ray of a small-calibre bullet in flight ever taken in the Soviet Union. We used the same simple technique for photographing a swiftly moving object that had been proposed at the end of the nineteenth century by the well-known Austrian physicist and philosopher Ernst Mach. The bullet would fly into the gap between two spheres connected in series into the circuit of a rectifier tube. The gap narrowed, was electrically breached, and the X-ray flash captured the bullet's image on film. We borrowed a small-calibre rifle and a dozen cartridges from a hunter friend of ours. The calibre of the bullets was 5·6 mm. For a bullet-trap we used a plywood box filled with lightly tamped sand. I can clearly remember those thrilling minutes when I first heard Avdeenko call out from the large cupboard which we used as a darkroom: 'We've got the bullet!' The image of the bullet flying at three hundred metres a second was quite distinct. Our standard kenotron was allowing us to make completely satisfactory X-rays of rapid processes.

In 1940, at the request of Academician E.A. Chudakov, director of the Institute of Machine Sciences of the USSR Academy of Sciences, our radiographic laboratory and all its staff

---

[7]Nizhnii Novgorod is the nearest major city to Arzamas-16 and is situated some two hundred kilometres to the north. In Soviet times it was renamed Gor'kii and is referred to as such elsewhere in this book. For security reasons, private telephone subscribers in Arzamas-16 had a Gor'kii telephone number.

were moved to his institute. Until the beginning of the war Avdeenko and I made great strides towards perfecting our technique for filming bullets and other rapidly moving objects using X-ray flashes. We perfected our synchronization technique so that we were getting images of bullets smashing both through wooden blocks and through other materials. Later we developed a technique for filming a bullet and an explosion in both visible light and X-rays simultaneously. This method allowed us to capture the motion of powder gases during afterburn, when the image conveyed by visible light is obscured either by the gases themselves or by splinters from the barrier.

Three years after the start of the war, when our daughter Irina started in the first class in School No. 43 on Kropotkinskaia Street, her teacher asked her the standard question: 'And what does your daddy do, Irina?' Irina thought for a second, then replied: 'He takes pictures of bullets'. Such an answer coming from an eight-year-old girl somewhat perplexed the teacher and she phoned the lab to check what sort of strange profession could involve photographing bullets.

Zina Azarkh

WE met when Zina was just finishing school. She was good at drawing and planned to enrol at the Institute of Architecture. The flash of mutual attraction we felt then quickly developed into a more profound feeling that filled and warmed our subsequent lives.

Walks together became indispensable. In those far-off days there existed the so-called Boulevard Ring which included our favourite Gogol Boulevard, along which the famous Tram A, the 'Annushka', used to go. However, our walks were not confined to Gogol Boulevard. Sometimes we would walk the whole length of Kropotkinskaia Street, cross the Garden Ring and Zubovskaia Square, where at that time there was a splendid boulevard, then venture still further along Great Pirogovskaia Street towards the Novodevichii Monastery. Sometimes, on the other hand, we would go up Gogol Boulevard to the Arbat, with its inimitable lanes. We would cover a good few kilometres, but we never felt tired. Our hearts would be filled with joy and happiness. Sometimes we would meet at the bakery next to the Kropotkinskii Gate. It was on this square that, until 1931, there stood the famous Cathedral of Christ the Saviour. It had been built by popular subscription in the second half of the nineteenth century in 'gratitude to God' for the victory over Napoleon and as a 'memorial for ages still to come'. The huge edifice, decorated with gold and mosaic and surrounded on all sides by broad staircases which, to the East, ran down to the Moscow River, was set on a hill. The interior furnishings of the cathedral were of great artistic value. They were the work of the best artists of the time—the sculptors P.K. Klodt, A.V. Logonovskii, N.A. Ramazanov, and F.P. Tolstoi, and the painters V.V. Vereshchagin, K.E. Makovskii, and V.I. Surikov. However, in 1931 it was decided to blow up the Cathedral and to build the Palace of Soviets in its place.[8]

During my final year at the Institute I was the chairman of the Union of Militant Atheists. At the behest of the Komsomol I would hold discussions with those who lived in the lanes adjacent to the cathedral. I tried to prove that the Palace of Soviets, designed by B.M. Iofan and crowned with a gigantic statue of Lenin, would be more meaningful and appropriate for our atheistic age. More than six decades have passed since then. Today I am ashamed to think that I took part in that act of barbarism.

They managed to lay the foundations for the Palace of Soviets just before the outbreak of war, but, however much they were reinforced with metal, they kept subsiding. The steel from them was eventually used for tanks. After the war, seeing how the foundation pits became filled with rainwater, they decided to build a swimming bath there. Through the force of inertia the metro station at the intersection of Gogol Boulevard and Kropotkinskaia Street was called 'Palace of Soviets' for many years to come.

On 16 October 1933 Zina and I went to see the opera *The Queen of Spades* at the Stanislavskii Theatre. Tchaikovsky's remarkable music has stayed with me ever since. Just three notes in sequence—do, re, mi—but what penetrating force these sounds possess in combination! The impression produced by the opera was so strong that we swore everlasting love to each other that very evening. Zina was sixteen. We agreed to get married as soon as she reached eighteen, and that we did.

---

[8]This church has now been completely reconstructed and was officially reopened on 12 June 1998.

In 1937 our daughter Irina was born. Zina was studying at the Institute of Architecture; my brother was also a student and my mother could not work because of illness. I became the sole breadwinner for the family and had to think about finding some supplementary income.

In 1941 the war came. Zina graduated from the Institute of Architecture. Along with the Academy of Sciences we were evacuated to Kazan. My eyesight was giving cause for concern. Slowly, but quite perceptibly, it was growing weaker. Zina did not want to give up her work—she found architecture enthralling. But the choice had been made—she had linked her life to mine. After a brief period of work in an architectural studio and a hospital, she came to work at our laboratory. From that time onward we worked together.

How can I possibly assess the role which Zina played in my destiny? She became my truest and most devoted friend. My problems became her problems, my successes—shared successes. The book you are now reading was written by us jointly.

It is difficult to write about someone who has been at your side constantly and has become something like a continuation of your own self. But I firmly believe that without Zina I would never have accomplished even half of what I have done.

In Kazan Irina developed tubercular bronchial adenitis after a bout of measles. Then in 1946 she fell ill with tubercular meningitis in Moscow. The illness was considered incurable. Clearly, undernourishment during the wartime years had played its part in her illness. Irina was the first child in Russia to be saved from tubercular meningitis. But the price of that victory was a total loss of hearing.

Zina spent about a year in the hospital. What did she feel when Irina literally nearly died in her arms several times? What amazing endurance and strength of will it must have taken to overcome this deadly illness and retain the capacity to love life and to rejoice in it.

And all the time the most terrible trial lay ahead—the illness and death of our son Sasha. His life was free from care. How much joy and happiness he brought to our family during the seventeen years he lived on this earth.

Sasha, Sasha...How terribly we failed you.

He was well prepared to take entrance exams in all the natural sciences, especially in physics and mathematics. In those years I was organizing classes in these subjects in Moscow. On Saturdays and Sundays the most talented children took part in these classes. The result was that out of thirteen children, all bar one got into institutes of higher learning. The unsuccessful one was our Sasha. This was 1966. At that time Jews were not accepted into institutes of higher learning. This unjust rejection led to depression, and a severe bout of influenza gave rise to a grave psychological disorder—hypothermal schizophrenia. This was diagnosed only in the last days of Sasha's life.

Our close friends did what they could to help. I will always remember the last week of Sasha's life—every day Lev Al'tshuler came to see us early in the morning to discuss a plan of action for the next twenty-four hours. Zina never left the hospital. Every day Sasha's temperature rose another degree and there was nothing that could be done about it. By the end of the week it had reached 42°C. Besides the high temperature another terrible symptom was developing—the nitrogen content of his blood was rising.

All the efforts of the intensive-care team to halt these processes ended in failure. His death agonies began during the night 28-29 October. For the last hour of his short life Sasha was conscious. I went up to him and ran my hand over his face. His brow was burning. All I can remember is the two words I said: 'Goodbye, son.' His ashes are buried in an old Moscow crematorium, situated next to the Donskoi monastery. The funerary niche is occupied by a large photograph of Sasha. The photograph, taken in 1966, shows a happy, open, boyish face.

Here I would like to include a passage from a letter written by a dear friend of ours, Marina Frantsevna Kovaleva. This is what she wrote to Zina one year on the occasion of her birthday:

Dearest Zinaida Matveevna,
    It's been a long time since we last met, and I want you to know that you live continually in my heart and in my memory.
    The very first impression I ever had of you was of an energetic, lively, even joyful woman, very straightforward and friendly. It seemed to me that life for you was easy, interesting and secure, that everything was working out for you exactly as you would have wished. Your interest in the theatre and theatre people seemed a pleasant bonus in a well regulated life filled to overflowing with the things you loved. It was only later that I learned about your misfortunes and Veniamin Aronovich's blindness. With every passing day, Zinaida Matveevna, you surprised me more and more.

Veniamin Aronovich very much wanted, very much needed, to feel a certain degree of independence, and you helped him in that, by trying to find all sorts of ways of effacing your 'guardianship', which was, nevertheless, constant. At lunchtime you would inconspicuously position a piece of bread near his hand while he felt for his spoon and fork on his own. If something was left on his plate you would say, as an apparently casual aside: 'Venia, South', and he would eat up what was left on the edge of the plate nearest to him. Over tea you would ask him: 'Would you like a biscuit or crackers?' And he would find what he wanted on the plate to the left of his teacup. Then you would say: 'Don't forget your medicine.' And he already knew that his tablets were on the table just to the right of his saucer. And all this took place against the general conversation at the table, in which Veniamin Aronovich participated most actively.

I suppose there must be a million other 'signals' devised by you which only Veniamin Aronovich can detect, thanks to the extraordinary sensitivity that exists between you. No-one could possibly keep count of all the trivial everyday items you have to remember in order for Veniamin Aronovich's life to run smoothly and calmly. The things to which he is accustomed have to lie or stand in their customary places in order for everything to keep to time—sleeping, eating and the early-morning drive to work. Veniamin Aronovich has no problems with transport—you are an excellent driver.

I shall never forget one conversation we had in hospital once, when his sight was not improving and he could not recognize your face...That time everything turned out all right for him.

An ordinary person could hardly cope with all the demands that are constantly made on you, yet you and Veniamin Aronovich give so much of your time and attention to others, to all kinds of truly good deeds, and you do it with such commitment, such inventiveness and such true creativeness.

I would willingly have termed your life 'heroic', but I know you would never have agreed with that.

In conclusion I would like to say that for many years now I've only been remembering what Zina looks like. I don't see the changes there must be in her appearance. And so I have an enormous advantage over everyone else. To me she will always be young.

MY EYES

MY eyes. There was something wrong with them. At dusk or in semi-darkness my sight was much worse than that of other people. It started in early childhood. I could scarcely make out the constellations in the night sky. I had trouble getting my bearings in the forest. During my first visit to the doctor he examined my eyes, and concluded: 'Night blindness. Drink cod-liver oil and get more vitamin A in your diet.'

I did everything he said, but my sight got no better. When I suddenly realized that I could lose my sight, the blow was unexpected. In the summer of 1936 Zina and I were given places in a holiday home in the little town of Kaliazin, near Uglich. A wooden barrack had been turned into a cinema with a portable projector. It is difficult now to say why, but the image they were getting on the screen was much dimmer than usual. Nevertheless, everyone else could see the image on the screen, whereas I not only could not see anything but could not even see where the screen was. It was that very evening in Kaliazin that I first thought: 'At this rate I could go blind.' That same evening, after the film show, I said to Zina: 'Think hard before you link your life with mine. Leading a blind man around is not the most pleasant of pastimes.' She put her arms round me: 'Whatever happens to you, I'll never leave you...' She has been true to this pledge ever since, remaining my constant friend and helper.

Gradually my sight impairment became more and more pronounced. Once, I bought an evening paper at the entrance to the metro and tried to read it.

'Why are you holding your paper upside down?' asked Lev Al'tshuler anxiously. 'Do you know, in this relatively dim light I can scarcely see the text. There's something wrong with my eyes.'

That spring Zina and I went to see the famous Moscow specialist, Professor M. Averbakh. He had treated Lenin in 1922 and 1923. When we met him, he was over seventy. After carefully examining my eyes, he pronounced his verdict:

'A rare form of pigmentary retinitis without any visible pigment in the retina. A serious illness; we don't know how to treat it. Your sight loss will get progressively worse.'

And, with a glance at Zina, he asked her: 'Are you expecting?' Receiving an affirmative answer, he thought for a moment and said: 'I wouldn't advise you to have children. Pigmentary retinitis is considered a hereditary disease.'

Before the war, in 1940, we made one more attempt to overcome my disease with the help of medicine. In that year I managed to be admitted to the Filatov Eye Clinic in Odessa. Filatov confirmed Averbakh's diagnosis and suggested treatment with biostimulators. This form of treatment had been developed at the clinic. After two months of treatment I was able to focus more sharply, but my field of vision remained narrow. At this time I was still able to read and write easily.

At this difficult period of my life Evgenii Bakhmetev lent me a great deal of moral support. He used to say: 'The loss of your sight is a great sorrow but I think you'll be able to overcome it and continue working. Your spatial awareness will develop and your concentration will improve. Homer was blind, so was Euler. Blindness did not stop them becoming great men.'

The irreversible fading of my vision forced me more and more to contemplate how my life would be from now on. There were two possibilities. The first was to turn myself into an invalid whose every thought is directed towards curing his blindness or waiting for a miracle or a discovery which would lead to a cure for tens of thousands of people who suffer from degeneration of the retina. The second was to adapt to the loss of my sight in such a way as to maximize my ability to work.

We chose the second path. Zina was at my side. She became my eyes.[9]

There is one 'positive' characteristic of pigmentary retinitis: the disease progresses slowly; vision fades over the course of decades. It's not like sudden blindness from a war wound or an accident. Nature in this case is more humane: it gives people the chance to adapt gradually to the loss of their sight. Time which would have been spent watching television or going to the cinema was freed up, visual impressions became briefer. The opportunity for inner concentration, so vital for scientific work, became a reality, my spatial awareness really did develop, and my powers

---

[9] A lightly fictionalized account of this period of Tsukerman's life can be found in Iurii Nagibin's story 'The Spring that Fell Silent' ('Zamolchavshaia vesna', 1982). See 'Bibliography'.

of memory improved. 'Inner vision' becomes intensely active in people who have gone blind, their hearing and touch become more acute. Their sense of direction is not impaired. The lack of visual information can be made up for by memory and imagination—mistakes can be kept to a minimum. They have their own particular ruses, which are gradually refined, in order that the need for help only arises in extreme circumstances. When I'm inventing or dreaming things up, I 'see' diagrams and designs down to the smallest details. Telling construction technicians about a piece of apparatus or a design I'd thought up did not present any difficulty. For a long time it helped me to draw sketches in chalk on black paper. Soon, however, this became impossible. Nevertheless, people who have worked with me for decades assert that when they are discussing new ideas and designs with me they are practically unaware of my blindness.

'I only discovered that Veniamin Aronovich was blind at one of our evening social functions when he rose from the auditorium to make some formal introductory remarks,' wrote Marina Frantsevna Kovaleva. 'And you, Zinaida Matveevna, remained where you were and watched him intently, quietly prompting him: "More to the left", at which he would make a slight adjustment to his course. Subsequently, people told me that Veniamin Aronovich knew exactly how many steps it was to the stairs leading to the stage, and how many stairs there were, so that he could do everything himself.

'Later, he himself demonstrated how he told the time, by taking out a pocket watch with no glass and feeling the position of the hands with his fingers. Soon I stopped being surprised at the way he dialled phone numbers, or typed, or related how he had "seen" the latest show and what he had liked in it and what he hadn't.'

After the war my focus and field of vision deteriorated noticeably. I could still read a typewritten text slowly in bright light. Up to 1953 I wrote my articles and reports myself, although I already could not read them. The manual habits of glass--working remained with me for a long time. Until 1952 I carried out experiments with explosive charges myself. From 1954 I needed assistance, especially during the hours of darkness. I acquired a typewriter and in two or three years I had mastered touch-typing. I could type as fast as a qualified typist. This should not cause any particular surprise. It is a well-known fact that qualified typists type without looking at the keyboard. Only such 'touch-typing' can ensure record typing speeds.

At work we encountered difficulties in focusing our X-ray chambers, when we had to make out a faint spot one millimetre in diameter on the fluorescent screen. Usually I demonstrated interference from monocrystals to the students. I knew, of course, that such instances of interference arrange themselves on elongated ellipses, but I had only seen them in photographs. The 'film speed' of my eyes was obviously insufficient to make out these same glowing spots on the fluorescent screen.

During lab sessions with the students, the whole group was subdivided into two groups. That day everything was going as usual: after explaining the main points of the methodology, I turned out the room lights and switched on the X-ray unit. After a few moments I said: 'Now that your eyes have had time to get accustomed to the dark, you can see greenish spots of varying intensity, arranged along elliptical trajectories.' And everyone could see them!

An hour later the experiment was repeated with the other group. But now, however much I tried, none of the students could see a thing. We turned on the lights, and opened the X-ray chamber. It turned out that Avdeenko had forgotten to put a monocrystal at the centre of the X-ray beam. So what did the first group actually see? Nothing, of course. Simply I had spoken with such conviction that they all 'saw something'.

There are a few more episodes connected with my loss of sight. In spring and summer I usually rode a bicycle to and from work. I was often detained at the lab until nightfall, in which case I had to take the tram home. But there were occasions when I would miscalculate how close it was to nightfall. The way home went down from the Pokrovskii Gate to Solianka Street. Once I knocked over an old woman in the darkness. I leaped off my bicycle, helped her to her feet, and kept repeating: 'Forgive me, please, I'm terribly sorry...I can't see very well.' When she had recovered her composure, the woman said: 'If you're blind, why are you riding a bike?' 'I won't any more, I promise,' I replied.

Something similar took place late in 1944. Our laboratory was then doing intensive studies of extremely sensitive explosive primers, such as lead azide and fulminate of mercury. The institute which produced these substances was in the area of the Nogatinskii Highway. The laboratory staff knew very well that transporting lead azide or similar substances by urban transport was against the law. Nevertheless there were occasions when we had to break the law. On such occasions it was usually me who transported the lead azide. We used a special box with shock-absorbers which Avdeenko had invented and constructed. My route took me first by tram to Vystavochnyi Lane and then by bus to Crimea Square. If I was detained until nightfall, seven-year-old Irina would wait for me at Park of Culture metro station. On this occasion the tram was delayed so long that it was dark by the time I reached Vystavochnyi Lane. I let a few buses go, since I couldn't make out their numbers. I had to turn to the other people waiting at the stop. People react very favourably to requests from a blind person. They helped me board the bus I needed, and half an hour later I was greeted with delight by little Irina. I decided against delivering the dangerous substance to the lab during the hours of darkness. The ten grammes of lead azide spent the night at our home, and were safely delivered to the test site at the institute the next morning.

When I told everyone at the lab what had happened, Al′t-shuler remarked: 'For a few hours you were a loose torpedo, weren't you?' During the last year of the war the seas and oceans were full of torpedoes which had missed their targets; they were christened 'loose torpedoes'. There were many incidents where military and merchant vessels were blown up by such torpedoes. Avdeenko, who was listening to this discussion, added, 'You know, there's another way you could deliver these primer materials. As soon as you're in the tram, just slip that little box of mine into the pocket of whoever's standing in front of you. And when you're ready to get off, carefully take it out again. Meanwhile, if it explodes, you can always say it wasn't your box...' Everyone laughed...It was not until the autumn of 1946 that our lab got its own car; from then on we used it to make our deliveries of lead azide and other sensitive explosive substances.

THE FIRST DAY OF THE WAR

SUNDAY, 22 June 1941 is a date engraved on everybody's memory.

Usually on Sundays I worked in the reading room of the Science Library of the Ministry of Fuel. Today it is still where it always was—on the corner of Nogin Street and Kitaiskii Passage. The day before I had ordered some books and journals from the stacks; now I had to look through them and make notes. On Saturday evening Zina had left for Leningrad to do her pre-diploma practical work, having completed her studies at the Moscow Institute of Architecture. Irina, then four, was staying at a *dacha* near Moscow. Since early on Sunday morning the radio at home had been switched on, playing some kind of music. Suddenly the network call sign sounded, followed by Levitan's voice: 'Attention! Attention! All Soviet radio stations are now broadcasting! Listen to the following important government announcement...'

This was Molotov's famous speech announcing the attack by Nazi Germany on the Soviet Union. He spoke as usual, stumbling slightly over certain consonants. The concluding words of this first war communiqué were: 'Our cause is just. The enemy will be smashed. Ours will be the victory!'

The news shocked me. However, this was long before *perestroika* and my behaviour in many areas continued to be stereotypical.

It was about one o'clock by the time I got to the reading room. Normally filled to overflowing with students, today it was empty except for a grey-haired little old man sitting pensively in one corner. I laid out the journals on the desk, but I couldn't keep my mind on my work. War! War! War!—kept hammering at my temples. After an hour I handed in my books and did not order any more. What had happened was slowly dawning on me. It was becoming clear that our whole existence was changing radically and that a new way of life was upon us.

I arrived home. The neighbours had come over, so had our relatives. Nobody could think straight but the majority thought that the war would be short. Voroshilov's recent speech was fresh in everyone's mind: 'We will beat the enemy on his own territory!' There was even a new term—'Voroshilov kilogrammes.' In one of his pre-war speeches Voroshilov had made a statistical comparison between our war munitions and those of other countries. Apparently we were far stronger than any capitalist country.

The First Month

AVDEENKO and I were excused army service: he because of a severe illness of the lungs, I because of my failing eyesight. On Monday 23 June there was practically nobody working in the institute. Everybody was talking about the war. Different dates were nominated for its end. People wanted to believe that it would last about three months, or six months at the worst.

At the end of 1940 and during the first half of 1941, the work we were doing at the radiographic laboratory had some bearing on military problems: we were developing superfast radiography. But then, during the first month of the war, this became irrelevant and our experiments in microsecond radiography were halted. We wanted to find something which might possibly quickly be of help at the front. At the request of the chief engineer at the hard alloys plant which had taken on the production of tungsten and tungsten carbide cores for armour-piercing shells, we began to deal with some of the technological problems involved in that production. Then, on 15 July, we got an official directive: all academic institutions were to prepare for emergency evacuation to Kazan. We set about packing our laboratory equipment, which was fairly bulky and heavy. We needed an entire freight waggon for it.

The blackout was introduced in Moscow literally the day after the declaration of war. Air-raid sirens sounded several times during the first month, but German planes were not yet penetrating as far as Moscow. The city was first bombed exactly a month after the beginning of military action, on the night of 21 July. Our nearest bomb shelter was the Park of Culture metro station. After the siren sounded I had time to get dressed and, taking with me a briefcase which I had got together that evening and which contained unpublished articles and other documents, I set off for the metro station together with all the other inhabitants of our block.

The first German bombing raid lasted the entire night. Near Crimea Square an anti-aircraft battery was firing without let-up. Apart from high-explosive shells up to two tons in weight, the Germans were also dropping a great many 'cigarette lighters'—incendiary bombs. They were also using flares, dropped on parachutes. The damage done during this first bombing raid was not very significant. However, it had a depressing effect. The Vakhtangov Theatre on the Arbat was destroyed. One of the theatres by the Nikitskii Gate was also destroyed and the monument to K.A. Timiriazev was damaged.

Our special train was scheduled to leave on 22 July. It set off about nine o'clock. We left behind the Kazan station, then the city with which all of us had so many links. However the train stopped about thirty kilometres from Moscow. That night the Germans made a second attack on Moscow, but our special train was a fair distance away from it. We could see aircraft with the swastika insignia when they were caught in the searchlight beams; we could see the anti-aircraft batteries firing at them. Soon the train started off again and in the afternoon of 23 July we arrived safely at our destination.

KAZAN

I STEPPED off the train and immediately saw a tall bearded man in a white suit, wearing the star of a Hero of the Soviet Union in his lapel. He was not difficult to recognize—the Vice-President of the Academy of Sciences and legendary polar explorer Academician Otto Iul'evich Shmidt. He had arrived in Kazan a day before our special train and was supervising the reception of the academic institutes and assigning accommodation in flats and student hostels to staff members of the Academy of Sciences. My family, along with Avdeenko's, was sent to one of the Kazan University hostels. It was a four-storey building in the Klykovka region on the outskirts of Kazan, about four kilometres from the university. At first we and the family of another staff member were given a large room on the third floor.

For a laboratory we were allocated a dingy room of about fifty square metres in a wing adjacent to the main three-storey building of the university. Legend had it that the famous Lobachevskii had worked in this very wing. It took us about a week to set up our basic radiographic equipment and begin work in our new location.

A week later Lev Al'tshuler's wife Marusia arrived with their two-year-old son Boris. Al'tshuler himself was serving at the front. She brought with her her sister Tania, who also had a small child. Marusia was taken on at the lab. She and her sister, together with their children, were housed in a student hostel on Bankovskaia Street, next to the university. Several other families were also living in the same room. The whole place had become a Noah's Ark.

From the time in early August when our radiographic equipment became operational, the question arose as to what our laboratory should do to bring about victory in the war as quickly as possible. At this time a new factor entered the picture. Hospitals with wounded soldiers were being evacuated to Kazan. Most of these hospitals had no radiographic equipment for examining the wounded. In Moscow Avdeenko and I had accumulated a certain amount of experience in assembling radiographic materials out of whatever materials happened to be available. We decided to use this experience in Kazan. We set up the first two X-ray machines using our own high-voltage transformers. Then, when the next evacuated hospital again turned out to have no radiographic equipment, we purloined a high-voltage transformer which belonged to Professor Dankov's laboratory and was standing idle in a corner of the main university building. With Shmidt's blessing we hauled the transformer out of the building late one evening and took it to the hospital on a commandeered lorry. The next day, when he discovered the loss of his transformer, Dankov played hell. Shmidt, however, came to our defence.

During the years of our evacuation in Kazan it took a great deal of inventiveness and improvisation to keep the X-ray equipment at the military and civilian hospitals running normally. At first the radiologists worked with diagnostic X-ray tubes. The tubes, however, soon gave out. Then we decided to replace them with high-voltage kenotrons—rectifying tubes. True, these kenotrons had a wide and indistinct focus. It was difficult to delineate caverns and cavities within the lungs with them. But shell fragments and other metallic foreign bodies showed up extremely well.

Some remarkable things happened during the time we supported Kazan's medical facilities. The X-ray machine at one of the military hospitals stopped working. We arrived and, as usual in such cases, the first check we made was on the oil level in the main high-voltage transformer. It turned out to be much lower than it should have been. We took the transformer apart—there it was: a high-voltage winding had broken through. We rewound the coil, topped up the oil, and the apparatus was working again.

The hospital director asked us: 'So where do you think the wretched oil went? You can see the transformer cabinet isn't leaking.'

A week later we were again summoned to the hospital. It turned out that the whole thing was the work of one of the cleaners. Somehow she had discovered that transformer oil burns very well in oil lamps, and, with the help of a length of rubber tubing, she had begun to siphon oil gradually out of the transformer tank. While the oil table was above the top of the windings, the apparatus worked. But when the oil level fell far enough to leave the windings exposed to the air, they began to break through into the cabinet.

There were plans to haul the cleaning lady before the courts. The laws operative in wartime threatened her with dire punishment. It took a great deal of effort to forestall a trial.

There was another 'oil story', but of a rather different kind. Before the war the French had begun to manufacture X-ray apparatus filled, not with transformer oil, but with castor oil. Compared with transformer oil, castor oil had two advantages. When it was absorbed by paper condensers, their volume increased by a factor of 1·5 to 1·7 on account of the greater dielectric permeability of castor oil. In addition castor oil has practically no corrosive effect on rubber. At that time good oil-resistant rubber was not being produced in the Soviet Union. At the beginning of 1941 we acquired through our suppliers Akademsnab some twenty kilogrammes of purified castor oil. An advance bottle of this oil was sent to Kazan. In 1942 we found out that superheated castor oil loses its laxative qualities. We decided to test it and fried some potato in castor oil. Nothing happened. In 1942, when food was especially scarce in Kazan, all the castor oil we had was used for cooking.

Aside from medical radiology, in Kazan we never for one moment forgot our basic speciality, technical radiology. In August and September several aircraft factories and an aircraft engine factory were evacuated to Kazan from Moscow. Before the war we already had had occasion to visit the radiology laboratory of the latter in Moscow. It was a simple matter to re-establish contact with the management of this factory. At the time there was an epidemic of faulty valves in aircraft engines. We had to set up mass radiographic inspection of those valves. By this time we had established contact with the cosmic ray laboratory of the Academy of Sciences Institute of Physics. This was situated in the main building of the university. Its staff—Oleg Vavilov, Vladimir Veksler, Nikolai Dobrotin, Il'ia Frank—were expert users of ionization chambers and other apparatus used to observe forms of ionizing radiation. We decided to pool our efforts—our lab would come up with an X-ray source and the cosmic ray laboratory would take care of its detection. The apparatus we built was given to the aircraft engine factory, which soon afterwards had one-hundred-percent radiographic inspection of its production of aircraft engine valves.

A similar piece of apparatus was built for the Izhevskii factory. Using mesothorium gamma rays, we found it was possible to make extremely precise bore measurements for sniper rifles.

The situation on all fronts was growing more complex. In mid-August the ring closed around Leningrad. German forces were driving furiously towards Moscow. In such conditions our work on constructing and installing diagnostic radiological equipment and seeing to its maintenance in military hospitals seemed of minor importance. So did our development of radiological and radioactive instruments for inspecting the components of aircraft engines and sniper rifles. We wanted something more—to offer real help to the front.

It is difficult to say now which of us—Avdeenko or myself—came up with the idea first: instead of throwing bottles of fuel mixture by hand, why not do it with gunpowder? The management of the institute gave our initiative its enthusiastic support.

That is how our laboratory got involved with an area that had nothing to do with either X-rays or mesothorium gamma rays. We worked in that area for about nine months, from October 1941 to August 1942.

THE BOTTLE-LAUNCHER

IT is common knowledge[10] that during the first months of the war our country was poorly equipped to defend itself against Hitler's panzer divisions. Anti-tank guns were not being delivered to the army in the required quantities. During these initial and most difficult stages of the war bottles charged with a flammable fuel mixture proved quite effective. They could be thrown from a distance of twenty to thirty metres. Once in contact with the

---

[10]Tsukerman is making a point here: until the advent of Gorbachev and his policy of *glasnost* (openness), such information was not well known to the general public in the Soviet Union. The same policy also made possible the publication of the Russian version of this book.

tank's armour, the fuel mixture would burst into flames, igniting the fuel tanks, and with that, the whole tank would burst into flames. Our bottle-launcher increased the effective range of this weapon to ninety or even a hundred metres.

After a series of experiments we developed and tested the following design. A special steel extension consisting of two tubes was fixed to the barrel of our tried-and-tested 1891-1930 version of the Mosina rifle. One of the tubes was a continuation of the existing barrel, the other widened to a diameter of 75mm. Into this second tube would be loaded, mortar-fashion, a glass vessel containing the fuel mixture and fitted with stabilizers and a primer. When the rifle bullet entered the barrel extension, powder gases passed into the mortar tube and expelled the glass vessel with great force. It would fly seventy-five to a hundred metres. The barrel extension allowed level fire at tanks combined with lofted trajectory fire of bottles of fuel mixture.

Work on the bottle-launcher was conducted at three institutes of the Academy of Sciences. In the laboratory of Natal'ia Alekseevna Bakh (the daughter of the well-known chemist Academician A.N. Bakh) a specially gelled fuel mixture was developed which would stick to the armour when the bottle shattered on impact.

Special primers were invented based on chromyl chloride. The fuel mixture, based on paraffin or petrol with a five-percent naphthalate gelling agent, was produced in the Institute of Colloid Electro-Chemistry. The primers were glass test-tubes fixed into the neck of the bottle in such a way as to smash on impact with the tank armour. When the liquid in the test-tube mixed with the petrol or paraffin mixture, it caused it to ignite.

Initial experiments conducted in December 1941 with conventional half-litre bottles demonstrated that the accuracy of the system was largely dictated by the geometric dimensions and weight of the bottle. Thus the need arose for special glass vessels shaped like mortar shells.

Thirty kilometres from Kazan was the 'Victory of Labour' glassworks. The senior foreman at the factory showed me and Avdeenko the rudiments of glassblowing bottle manufacture. The Kazan glassblowers agreed to produce the glass mines for us on condition that we delivered split-pattern metal moulds to them. Following Diesel's celebrated maxim that 'Engineers can do anything', we mastered the production of such moulds. Later, when we were already working in Arzamas-16, this experience came in very handy.

Soon after we sorted out production of bottles at the 'Victory of Labour' factory, a new problem arose: how to deliver them to Kazan. The suburban trains were infrequent, did not keep to the timetable, and were overcrowded. In the city a number of cases of typhus, that perpetual accompaniment of almost every war, had been recorded. There were fears of an epidemic. Help came from an unexpected quarter, in the shape of two young women— Lidia Vasil'evna Kurnosova and Zina. Both fine skiers, they decided to do the whole journey on skis, carrying rucksacks in order to deliver what we needed so badly. As I equipped this female brigade for its mission, I gave then a hundred grammes of spirit between the two of them and two onions each. At the factory they were given hot soup and horsemeat. The women coped magnificently with the problem of delivering the bottles. Their 'ski outing' took them forty-eight hours.

At that time women formed the backbone of our laboratory. Apart from Zina, Al'tshuler's wife Marusia and Avdeenko's wife Liudmila Stepanovna were working there. The women did the necessary measurements of our glass mortar shells and took an active part in test-firing them.

LIDA

I WOULD like to say something about Lidia Vasil'evna Kurnosova. Her husband was Oleg Vavilov, son of the famous geneticist Academician Nikolai Ivanovich Vavilov. They arrived in Kazan in late July 1941. Oleg was on the staff of the cosmic ray laboratory at the Institute of Physics. He and I worked closely together on X-ray and gamma ray thickness gauges. Together with Il'ia Mikhailovich Frank (now an Academician), he was co-inventor and chief constructor of the ionization chambers used to measure barrel-wall thickness in sniper rifles at the military plants which had been evacuated to Kazan.

Lidia, or Lida as she was known, had just graduated from the physics department of Moscow University and became a researcher in our institute. Her main job was research into the durability of joints in high-calibre machine guns and automatic aircraft gunnery. Strain gauges were attached to the joints. Under tensile and compressive stresses the gauges would deform, altering their electrical resistance. These variations in resistance provided a measure of the forces acting on the gauges. Lidia coped with this apparently unfeminine work with great skill.

Test firing was carried out in a special shooting gallery. The properties of dozens of components were studied there. The results were quickly passed on to the arms manufacturers, who took steps to reinforce components and joints wherever necessary. This pretty, dark-haired and dark-eyed young woman quickly gained the highest respect in the laboratory.

Our two families became friends. We observed holidays and other important dates together. These young, talented people, full of joie de vivre, were very attractive.

In Autumn 1943 we learned that Oleg's father, Nikolai Ivanovich Vavilov, who had been arrested in 1940 in the course of Lysenko's anti-genetics campaign, was being held in prison in Saratov. Oleg did everything he could to visit his father, but bureaucratic obstruction over a pass and permission to visit prevented him doing so immediately. When Oleg arrived at the Saratov NKVD, he was told that his father had died of malnutrition shortly before. He was unable to discover the whereabouts of his father's grave. Only recently there was a brief note in the press about an eyewitness who chanced to observe the burial of Vavilov in a common grave in the Saratov cemetery.

Oleg and Lida left Kazan and returned to Moscow six months ahead of our group. Oleg completed his dissertation on cosmic ray research and successfully defended it in January 1946. Zina and I attended the dinner party held to celebrate the event. I remember Vladimir Veksler proposing toasts in honour of the new doctor of physics and mathematics. However, life took a different course. Something dreadful happened...Soon after defending his dissertation, Oleg left for the Dombai Valley where the Alibek Glacier, familiar to every physicist who studies cosmic rays, begins. Oleg was a very good mountaineer. He left...and did not return. The situation turned out to be similar to the one described by Vladimir Vysotskii.[11] A snowstorm started, which soon turned into a blizzard. Oleg fell from the cliff face, and his

---

[11] A reference to Vysotskii's 'Song about a Friend' ('Pesnia o druge', 1966). The song recommends taking a friend, whose loyalty seems questionable, on a mountain-climbing expedition and yoking him to you by a single safety rope. This will demonstrate 'who he really is'.

companion, whom he regarded as a friend, did not go to his aid but returned to base camp with the news that Oleg had perished. This came like a bolt from the blue. Lida decided to go immediately to Dombai and organize a search party herself. The first expedition, in late February 1946, ended in failure—there was too much snow in the mountains. The second expedition, organized in the summer of the same year, used experienced mountaineers and equipment capable of detecting small metal objects. This time Lida herself found Oleg's body. He had crawled a few dozen metres from where he had fallen. That means he survived the fall. A large pocket watch, a gift from his father, showed two o'clock. It had probably stopped when he fell.

Oleg was buried in the Dombai Valley. There is a small cemetery there for victims of climbing accidents. A decade later, during a car trip we were making from Tiberd to Dombai, we stopped at this modest grave. Lida had requested the following inscription for Oleg's gravestone:

'Here, in February 1946, perished Oleg Vavilov, a talented scientist and the nearest and dearest person in my life.
Lidia Kurnosova-Vavilova'

We took a few photographs of the gravestone. And every time I re-read these sorrowful words, I was reminded of another inscription, made on Mount David in Tbilisi by another young, beautiful, dark-haired woman—Nina Chavchavadze on the grave of her husband Aleksandr Sergeevich Griboedov.

'Your wit and your deeds are immortal in Russian memory, but why did my love survive you?
To my unforgettable one, from his Nina'

More than a hundred years separate these two inscriptions, and yet some elusive quality seems to unite them...

Oleg's unexpected death changed all Lida's plans. She transferred to the Department of Cosmic Rays of the Institute of Physics of the Academy of Sciences. She considered herself in duty bound to continue Oleg's work. She worked doggedly and selflessly. In 1954 she successfully defended her dissertation and received her doctorate in physics and mathematics. The theme of her dissertation was 'The Compton effect at a photon energy level of 250 mega-electron volts'. In October 1957 our country launched the world's first orbital satellite. Subsequently launched satellites were equipped with instruments for recording cosmic

31

rays which had been developed by a research group headed by Lidia Kurnosova.

In Autumn 1957 Lidia gave a report on satellite-based cosmic ray research at the International Congress of Astronauts in Barcelona. The organizing committee of the Congress handed out mock passports for the first manned flights to the moon to all the delegates. Passport no. 2 went to Lida.

14 July 1942

The days we spent preparing and conducting test firings of our glass vessels were especially memorable. Measurements and quality control took up a great deal of time. Often these operations would continue long after midnight. In the morning we would take our bottles and primer assemblies and set out for the training range, located about ten kilometres from Kazan. There we would conduct test firings, using captured German Tiger and Panzer tanks as targets. These tests in Kazan went well. It was decided to send a team with the apparatus and a supply of ammunition to one of the testing grounds near Moscow. Marusia and I took part in this trip.

Test firing was conducted on the testing ground at Solnechogorsk near Moscow. The date is easy to remember—14 July—Bastille Day, only the year was 1942. We had two boxes of bottles for test firing, with twenty-four bottles in each. While negotiations were under way over the order of test procedures, I began to fit primers to the bottles. The first twenty-four bottles were successfully primed and made ready for test firing. But when I began to fit a primer to the twenty-fifth bottle, a bright flame suddenly leaped up. The bottle instantly shattered in my hands and half a kilo of burning fuel mixture ended up in my lap. I was cut off from everyone and everything by tongues of flame, which licked my face and arms. The thought flashed through my mind that I was now playing the part of a burning tank and could be devoured by the flames. But my mind was working clearly: I had to get clear of the remaining bottles, which contained another thirty kilogrammes of petroleum mixture. Marusia tried to help me to undo the buckle of my belt. But it was no good—the fire had covered the buckle. 'Can this be the end?'—I thought. Images of my mother, my wife and our five-year-old daughter passed instantly before me. But suddenly, miraculously, two Red Army men appeared from nowhere. They ripped apart the burning belt.

One seized me by the arms, while the other pulled off my trousers and the burning fuel. They smothered my underpants with sand. In a matter of moments the fire was completely extinguished.

In the medical unit at the testing site the doctor found third-degree burns on my arms and knees and second-degree burns on my face. An ambulance took me to Moscow, to the military hospital on Basmannaia Street.

It was not difficult to determine the cause of the fire. One of the primers turned out to be longer than prescribed. When I inserted it in the bottle, it broke. The primer liquid combined with the fuel mixture, which burst into flames.

After I had been in the hospital for five days Zina arrived. It had taken two days for her to get from Kazan in a goods waggon. My temperature was still high, more than thirty-nine degrees. I was delirious, but I recognized Zina.

The Germans were driving relentlessly towards Stalingrad. There were almost no bombing raids on Moscow but air-raid sirens were still heard, although rarely. It's a wretched thing to have to deal with—air-raid sirens and hospital, with your arms and legs bandaged up so you can't take a single step. Luckily, however, they never had to evacuate the wounded.

The doctor reckoned it would take at least three months to recover from burns like mine. Lying about in hospital for such a length of time formed no part of my plans. In late August the doctor consented to discharging me early. Perhaps Zina's selfless care had had its effect. She would arrive at the hospital at nine o'clock every morning as if she were reporting for work. She took care not only of me, but also of the other wounded in my ward.

When I was discharged, the skin on the wrists and fingers of both hands was so tender that a handshake would make it peel off and bleed. Not only could I not carry anything heavy, I also could not play the piano. Nevertheless, I had to continue what I had started. The mortar designer B.I. Shavyrin had taken an interest in our bottle-launcher. His design team and testing ground were situated at Golutvin, near Moscow. We decided, without further ado, to go to see him with our remaining box of bottles.

X-RAY OF AN EXPLOSION

WE spent several days with Shavyrin. There an important event took place which would shape my own destiny and the destiny of our laboratory for many years to come. Shavyrin told us of an event which had disturbed many military designers. During one of their counter-attacks towards Tikhvin our soldiers had seized a depot of German munitions, in which they discovered shells of a new type. They called them 'armour-burning' shells. In the front part of these shells there was a conical or hemispherical cavity from 1·5 mm to 2 mm thick. Although this forward cavity meant that the volume of the 'armour-burning' shell was less than the volume of explosive carried in a conventional shell, it could still pierce armour three to four times thicker than that which a conventional shell of the same calibre could pierce. The hole it left in the armour really did resemble a burn-through in flammable material. No-one could understand how these shells worked.

All day long I couldn't get this 'armour-burning' phenomenon out of my mind. Then, that night, the following happened.

Zina and I had taken a room on the second floor of a small two-storey hotel attached to a factory. After test-firing the bottles we returned to our room about nine o'clock in the evening. We had not yet fallen asleep when the air-raid siren sounded. The Germans were bombing Kolomna and the warehouses situated five or six kilometres from Golutvin. We did not know where the air-raid shelter was and we did not want to leave the hotel. I was still mulling over those armour-burning shells. Suddenly, like a flash of lightning, a solution came to me which drove everything else from my mind: we had to use X-rays to film the explosion of one of those shells. The X-ray would allow us to see the behaviour of the metal casing and to understand its purpose.

When they had dropped their bomb loads, the Germans flew off and I spent all night thinking over the technology needed to film the phenomena produced by the explosion. Naturally, this would be more complicated than filming a shell in free flight, when there is no shock wave to deal with and it is easy to protect the X-ray plate from damage. But, if we protected sufficiently the cassette housing the X-ray plate, it would solve the problem of filming explosions.

In the morning we quickly did a series of test firings of bottles and by the evening we were back in Moscow. I knew that at that time one of our leading specialists in explosion and detonation phenomena, Iulii Borisovich Khariton, was living in Moscow, in the hostel of the Academy of Sciences. In January 1942 Khariton, Ia.B. Zel'dovich and I had discussed the problems of expelling our glass shells with powder gases. I had to see Khariton.

In the early morning of 25 August Khariton and I were to meet in the Academy Hostel in Neskuchnyi Garden. This is right next to the building which now houses the Praesidium of the Academy of Sciences. Khariton heard me out, and asked a few questions: 'I strongly advise you to put off everything else you may be working on and to concentrate on this very promising methodology. We have a staff member of our laboratory working in Kazan, Aleksandr Fedorovich Beliaev. I'll write him a letter, and he'll help you master explosives experimentation.'

He thought for a while, then added: 'If you like, go and consult the deputy chairman of the Chief Directorate of Artillery, Lieutenant-General Konstantin Konstantinovich Snitko. I have a feeling he'll also give you his support.'

That afternoon I went to the Chief Directorate of Artillery. The general received me attentively. I told him about our little epic with the bottles and about the possibility of using radiographic methods to study the mechanism of shell explosions. General Snitko thought for a few minutes, and said: 'I think you should abandon your bottles. That's yesterday's news, last year's news if you like. Now our factories are supplying the front with ever-increasing quantities of anti-tank guns and armour-piercing rifles. But if you can get to the bottom of the mechanism of this cumulative ammunition (this is what they had already started to call armour-burning shells), that would be a great help both to the munitions factories and to the front.'

In late August 1942 Zina and I returned to Kazan. I had struck up a close and friendly working relationship with A.F. Beliaev. He kept me supplied not only with technical literature, but also with detonator capsules, detonator cable, and explosive powders—tetryl, hexogen and PETN. The radiographic laboratory changed course once again. Incidentally, we were now closer to our pre-war areas of research—we were applying X-rays to the study of explosive phenomena.

People often ask me: whatever happened to those beloved bottles of yours? I have no regrets at all about what I did. You should only regret the things you didn't do. All that tinkering with bottles gave us a good grounding: we learned the fundamentals of ballistics, internal and external; we learned how to work with glass. Later on this experience stood us in very good stead. Furthermore we arrived at the radiography of explosions, in essence, as a result of our work with those bottles. Who knows, if we had not gone to Golutvin, if there had not been a German air-raid with the sleepless night it entailed, we might never have hit on the radiographic method for studying cumulative explosions and cumulative charges.[12]

It took about three months to effect the transition to the radiography of explosive processes. Our laboratory in Kazan had an adjacent vestibule. It was decided to convert that vestibule into an explosion chamber. We boarded up the windows in the vestibule and reinforced them with sandbags. The laboratory's sole remaining transformer was adapted for charging capacitors.

Shielding the radiation source and the X-ray plate from explosion damage was quite a complicated problem. Still more difficult was the achievement of synchronization of the X-ray flash with the necessary phase of the explosion process. At the end of December 1942 we got our first X-rays of the explosion of lengths of fulminate-of-mercury detonating cable. The presence of the fulminate of mercury in the composition of the explosive substance enhanced the contrast in the X-ray image.

In 1943 I successfully defended my dissertation for the degree of Doctor of Technical Sciences. My official examiners—V.I. Veksler and A.I. Shal'nikov—were unanimous in noting the particular importance of microsecond radiography for the study of explosion and detonation phenomena.

In March 1943 we learned how to obtain tolerably good image resolution in X-rays of explosion phenomena. At last the role of the metallic casing of the cumulative cavity in the German armour-burning *Panzerfaust* bazookas was revealed. It turned out that, in the course of exploding, a high-speed jet of molten metal arises along the charge axis, the formation and development of which were clearly visible on our X-rays. It is this which gives the cumulative charge its armour-piercing capability. Academician M.A. Lavrent'ev came to the same conclusion on the basis of

---

[12]Russian: *kumuliativnyi zariad*. Also translated as 'shaped charge' and 'hollow charge'. I am indebted to Dimitry B. Sergay for the information that Tsukerman's description of these charges omits a crucial principle involved in their functioning.

theoretical hypotheses. The ideas which he developed led to a definitive theory of cumulative explosions.

By mid-1943 the situation on all fronts had begun to change. The battle of Stalingrad had been won. The rapid advance in the Kursk salient had begun. Some academic institutes returned to Moscow. For us, this was impossible. Organizing the country's first flash-radiography research laboratory required a great expenditure of time and effort and it would not have been sensible to start everything again from scratch in Moscow.

During the second half of 1942 Lev Al'tshuler returned to our laboratory. Because of the successes at the fronts, the government allowed the Academy of Sciences to recall a thousand of the most highly qualified scientists. Al'tshuler took up super-high-speed radiographic structural analysis. In spring 1943 he successfully defended his doctoral dissertation.

I remember one seminar in the Leningrad Institute of Physics and Technology. That institute, like the Physics Institute of the Academy of Sciences, occupied several rooms in the main building of Kazan University. The director of the institute, the renowned physicist Abram Fedorovich Ioffe, was chairing the seminar. I explained the general principles of X-raying shells in flight and of explosion phenomena, and showed some X-ray plates. Al'tshuler showed us his light-proof chamber for radiographic structural analysis of fleeting processes and the first X-rays of polycrystalline specimens, taken at an exposure speed of less than a millionth of a second.

Ioffe, who was always a great enthusiast, gave a very high assessment of these studies. He said that this material was almost as important for the field of X-ray science as the discovery of the North Pole had been for geography. It is curious that what impressed him most was the study of the structure of metals using microsecond radiographic exposures. The filming of gunshots and explosions became widespread during the war and in the post-war years. On the other hand, the structural analysis of materials in microsecond and nanosecond durations is only now, over forty years later, beginning to come into its own. This kind of research turned out to be far more complicated than the direct illumination of quickly-moving objects.

In May 1943 Ia.B. Zel'dovich made a series of X-rays of exploding models of cumulative shells and took them to Moscow to show Khariton. Soon afterwards I received a letter from

Khariton. During the war and evacuation the letter was lost, but I can remember the contents. It began with the words:

> Dear Comrade Tsukerman. Forgive me that I do not know your name and patronymic. I showed your unique X-rays to the People's Commissar for Munitions, B.L. Vannikov. They made a great impression on him, as they had done on me. We agreed that you should give a report to the Board of the People's Commissariat for Munitions in July or August...

This was followed by remarks on what was required in the report and in the demonstration materials.

This report, which was momentous not only for me, but for our whole laboratory, was given in mid-August 1943. It went well. A resolution was passed to begin to organize two flash-radiography laboratories in Moscow: one in an institute dealing with explosive substances, and one in an institute which designed and tested shells.

WORK AND LIFE WITHOUT ELECTRICITY

I HAVE already said that aircraft factories and aircraft engine factories were evacuated to Kazan. These were energy-intensive industries, and the energy supplies in Kazan were unable to meet the increased demand for electricity. But there was a particular need for aircraft at the front, and their production could not be halted for a moment. Accordingly, during the winter months, when demand for electricity was especially high, the whole city would be disconnected from the grid for two, or even three weeks. Was it possible to work in conditions like that? It was. During a physics practical I spotted a disc-type Whitehead electrostatic machine, which had seen better days. I knew that by turning the handle of such a machine you could charge up Leiden jars—that's what old-fashioned glass capacitors used to be called—up to 100 kV and higher. It was not difficult to arrange for this machine to be used temporarily for flash-radiography and for it to be carried by hand to our own 'lair'. We connected the Whitehead machine's high-voltage lead to our capacitor. And sure enough, after three or four minutes cranking of the handle we had our 0·005 micro-farad capacitor charged to sixty kilovolts.

We switched the filament voltage supply for the kenotron, which was our X-ray source for microsecond flashes, over to our battery feed, so that we were not dependent on the mains supply. The problem of filming rapid processes was solved.

The darkroom also proved easy to manage. At first they wanted to make it out of plywood, but it was very difficult to get hold of sheets of plywood. A lucky chance came to our aid. In one of the local furniture stores a large quantity of plywood seats for chairs was discovered. We lost no time in acquiring ninety such seats, each one 50cm square. In a corner of the laboratory a framework was made, to which the seats were attached. In addition, we glued black paper over all the gaps and cracks. The result was a perfectly good darkroom, four square metres in area. The cheerful black and white squares on the seats even added a certain something to the decor of our laboratory. For a source of red light we used one small opening, covered with red glass. For night work we used a lantern with a wick.

The greatest difficulties arose with equipment which required vacuum conditions. We had mercury diffusion pumps in our lab which could pump out air to the necessary degree of rarefaction, but for them to work, you had to have the mercury reservoir warmed up and a source of water for cooling. And the water supply, like the electricity supply, was erratic. However, we coped with this easily enough. We set up a reservoir water tank with a capacity of three cubic metres, and when the water supply was cut off, we switched to our own system. Heating up the mercury reservoir, on the other hand, called for some ingenuity. Avdeenko's 'golden hands' came to our aid. By then he had mastered the construction of *burzhuiki*. That was the name that had been given since the Civil War to small iron stoves with a pipe that led directly out of the *fortochka*. Avdeenko rigged up a very small *burzhuika* with the help of which we could warm up our diffusion pump. A couple of dozen pieces of kindling would suffice for the whole working day. It became possible to continue our experiments with demountable vacuum tubes without a mains electricity supply.

The lack of electricity also made for additional difficulties in the hostel on Klykovka Street. I managed to get hold of a hundred or so bricks; I worked out how to arrange them, and a real stove was set up in our room on the third storey. It both kept us warm and served as a cooker. Dinner—mostly pea *kasha*— would be cooked on it, not only by us, but also by our neighbours. Thus this stove more than once rescued our immediate neighbours, the mathematicians Liudmila Vsevolodovna Keldysh and her husband P.S. Novikov.

Avdeenko made a large L-shaped stovepipe out of tin. We had to set it up with its outer end at the level of the roof ridge. Here a problem arose: who would set up this fairly sizeable apparatus on the roof of a four-storey building? I was no good because of my eyesight, neither was Avdeenko. Zina volunteered. We tied a safety rope to her made of bed sheets folded over four times. She climbed out of the dormer window and onto the roof, and successfully fixed the stovepipe firmly in place.

The history of *burzhuiki* in flats would not be complete if I did not describe one more 'invention'. A serious shortcoming of these small stoves was their unsuitability for windy conditions. Even a light wind would blow smoke into the room and put out the stove. In order to overcome this I suggested that we fix a Venturi tube on the upper section. This consists of two hollow truncated cones arranged base to base. Normally the stovepipe is attached to the axis of the Venturi tube, at its centre, and the resultant rarefaction increases the outflow of smoke from the stove into the atmosphere.

Avdeenko and I tested this device for increasing the draw of the stove during windy weather and were convinced of its effectiveness. Unfortunately, the idea was not fully realized. We should have linked the Venturi tube to a weather-vane in such a way that the axis of the pipe always corresponded with the direction of the wind. But we did not have time to make such a device. Moreover, even the fixed Venturi pipe produced a positive effect. While the neighbouring room, which had a conventional *burzhuika*, filled up with smoke in windy weather, there would be a complete absence of smoke in the rooms where our families were using the stoves.

When I remember these distant years, I can't help thinking that it was a remarkable school which taught us a great deal.

Soon after the war some captured reports from the Kaiser Wilhelm Hochschule fell into my hands. Tidy yellow folders with the words 'ganz geheim' (top secret) in the upper right hand corner. At the end of the text the obligatory 'Heil Hitler' before the signature. It is curious that the Germans had begun radiography of explosion phenomena, in particular, of models of cumulative charges, at practically the same time as the Soviet Union. The head of the German research programme was a certain Rudi Schall. After the war he lived in West Germany, where he published a number of articles on flash-radiography. But he, of course, did not have to develop high voltages by cranking the handle of an electrostatic machine, and the diffusion pumps of his demountable vacuum tubes were heated by gas burners or electricity, not by pieces of kindling.

We spent the winter of 1943-44 working in Kazan. Then, in early spring 1944 we once again loaded all our equipment into a freight waggon and, together with our families, we left for Moscow, and home.

MOSCOW 1944-47

IN Moscow we immediately set to work assembling and building our radiographic apparatus for explosion phenomena work. I also worked on organizing and assembling similar installations in two Moscow institutes involved in the study and design of munitions. In all equipment of this generation the source of microsecond X-ray flashes was still the kenotron with its cathode capable of sustaining short-term overheats.

Among the events of the last years of the war, three stand out in my memory: the death of my mother, my paper for P.L. Kapitsa's seminar and the granting of awards and medals to a large group of scientists and technicians.

MOTHER, DON'T DIE

> Whoever reads this,
> Bring lilies of the valley quickly to your mother.
> For mine, it's too late; bring them to your own.
> A. Voznesenskii[13]

MOTHER! What a great and tender word that is. My memory holds profound love and gratitude towards her. I can easily recall her beloved form, her kind, unique face. I only have to go to the piano and play one of the Ukrainian songs of which she knew so many. We knew from photographs how beautiful she had been in her youth. She was only thirty-four when our father died from inflammation of the kidneys. After his death, many sought her

---

[13]From the poem 'Mother' ('Mat'', 1983) by Andrei Voznesenskii.

hand in marriage, but in vain. She considered that a stepfather in the house was worse than no father at all. She devoted herself entirely to her children.

She possessed excellent musical pitch, which my brother and I inherited to a certain degree. She had another quality of no little importance—a memory for music and verse. She could recite Lermontov's *Song of Tsar Ivan Vasil'evich*... and *The Demon* by heart. She had a good memory for figures and formulae. In Petersburg she had graduated from the Women's Higher Education Courses and had a superb knowledge of geometry, algebra and trigonometry. After the death of her husband she worked and supported the family single-handed.

She had her first stroke before the war, in 1935. She stopped working after it. She had her last stroke in Moscow on the night of 28 March 1944. For several days before this I had been thinking very hard about how to obtain not a single 'still' or 'frame' but a series of chronologically consecutive radiographs of the process of an explosion. I had been working day and night on this. This is the only way you can think up something new. At about two in the morning came the flash of insight for which I'd been waiting so long. We had to use detonating cable as a means of predetermining the time intervals between frames. Then we could get several frames on a single X-ray.

The problem was almost solved when a strange sound came from the next room. I ran there at once. Mother was lying on the floor next to the light switch, two metres from her bed. I could detect no breathing. Could this be the end? Mother, don't die! Emergency first aid was of no avail. The ambulance arrived and the doctor pronounced her dead.

How often I berated myself subsequently. I had intended to move the switch closer to her bed. A small job, but I never got round to it. For that matter, her second stroke, which had happened in Kazan, was also my fault. I'd found a length of old wire with silk insulation which I'd used to connect the electric socket to the mains. I knew perfectly well that old wires are unreliable, and that if overloaded they can catch fire. That's what happened. Within a minute the whole wire was burning. The fire was quickly put out, but Mother had been frightened, and that very evening she suffered her second stroke. I should have paid much more attention to her. How often it is only after the death of someone close that we begin to understand what an irreparable loss we've suffered. How little we did for them. The pangs of conscience torment you for the rest of your life, but they come too late. Once again I'm tormented by the realization that I didn't move the light switch when I could...Only a few metres...

Kapitsa's Seminar

MY paper for P.L. Kapitsa's seminar turned out to be something of a historical milestone. At that time the famous 'Kapitsa Wednesdays' were drawing crowds of physicists from Moscow, Leningrad, and other cities. The seminar usually consisted of two papers, each lasting forty to fifty minutes.

On 8 March 1944 the first paper was given by Iulii Khariton. It was on the mechanisms of explosive reactions. The second paper was given by me and was on the subject of flash radiography of explosions. Petr Leonidovich Kapitsa was in the chair. This was my first encounter with him. I was struck by his engineer's grasp of subject matter. I remember, too, his unexpectedly high voice. He pronounced the Russian word *kondensator* like its English equivalent 'condenser'.

There were present at the seminar the well-known physicists A.F. Ioffe, L.D. Landau, I.E. Tamm, N.N. Semenov, Ia.B. Zel'dovich, S.I. Vavilov, and I.V. Obreimov. My paper generated a great deal of interest. Many present knew that it had been nominated for a State Prize (known at that time as a Stalin Prize). Immediately after the paper Lev Davidovich Landau came up to me and congratulated me on an excellent piece of work. A.I. Shal'nikov asked me: 'You described the X-ray images of explosions with such flair that it seemed to me you must have started seeing better in Moscow. Is that the case?' 'No, Aleksandr Iosifovich, pigmentary retinitis is an irreversible eye disease. It's simply that I've shown these X-rays many times and I'm familiar with them. Besides, I've worked out my own little tricks. You see these small cuts in the borders of the X-rays? I can feel them easily with my fingertips. They are an almost infallible guide, enabling me to show the audience those parts of the X-rays which merit particular attention.' 'That's a fine piece of work you have done,' Shal'nikov concluded.

The First Award

On 5 October 1944 the newspapers published a decree of the Praesidium of the Supreme Soviet of the USSR announcing the awarding of orders and medals to a large group of scientists and engineers who had been involved during the war with the research and development of various munitions. Mikhail Ivanovich Kalinin presented the awards in the Kremlin. We were warned in advance: 'Comrades, take care not to shake Mikhail Ivanovich's hand too vigorously, please. Remember, he's sixty-nine, after all.' All the same, he looked younger than that. In congratulating us, he found a few kind and encouraging words for each of us in turn. In the list of award-winners there were a number of people from the Academy of Sciences. Khariton was awarded the Order of the Red Star, I was awarded the Medal of Honour. For both of us these were our first government awards.

The main event of 1945 was, of course, our victory over Nazi Germany. Victory Day, like the first day of the war, was unforgettable. Since the morning of that day, 9 May, there had been rumours that Stalin would make a speech that evening. Irina and I set out for Red Square at five p.m. It was overflowing with people. The feeling of a great holiday, of universal jubilation, never left us. Total strangers were weeping together, laughing, kissing...We unexpectedly ran into the head of our laboratory, Nikolai Petrovich Raevskii. He scooped up eight-year-old Irina and sat her on his shoulders. The future seemed bright and festive. People were discussing their immediate plans. Everyone thought that this war would be the last one. No-one knew that in these May days the United States was preparing to test its superbomb.

Visit to Leningrad

Two weeks after Victory Day Zina and I went to Leningrad to visit one of the small arms institutes. We brought with us a rifle (the 1891–1930 model), ten cartridges and our synchronization apparatus. In forty-eight hours, using some additional materials, we had put together a flash radiographic installation and adjusted the synchronization. The paper went well. We not only managed to photograph a bullet in free flight, but also obtained a very striking shot of a bullet penetrating a light bulb in the instant before the glass shattered.

Leningrad was still a military city. We walked along Nevskii Prospect. On the buildings there were still notices: 'During artillery shelling it is dangerous to walk along this side of the street'. There were many buildings almost totally destroyed. The walls of some of the residential buildings had been completely stripped away, yet in the rooms objects from ordinary life had been preserved: a table still standing, a bed, an open piano. Here and there chandeliers dangled helplessly. We shuddered to think what had befallen the people who once lived in these flats now exposed to the four winds.

And then, in that half-destroyed city we came across an unexpected advertisement: a concert performance of Grieg's *Peer Gynt* at the Philharmonia. Grieg is a favourite composer of mine. That evening Zina and I were wholly captivated by his music, surrounded by a fantastic world consisting of the wild Norwegian landscape, wizards and trolls. But when the orchestra began to play 'The Death of Åse', memories of my recent loss again began to torment me. 'I can't listen to any more,' I whispered to Zina, and we quietly left the hall without waiting to hear the wonderful, enchanting, always thrilling 'Song of Solveig'...The White Nights were only just beginning, and that enhanced the mood the music had created. On the Anichkov Bridge they were just finishing the re-erection of Klodt's magnificent horses.

AUGUST 1945

ON that August morning of 1945 I woke early, as usual. I groped for the vernier dial on the wall, which was wired to the loudspeaker of the radio broadcasting network. I gave the dial a slight twist. The sound of the Kremlin chimes rang out. The announcer began to read a TASS communiqué. It appeared that the previous day President Harry Truman had made a speech on national radio. He had announced that at eight o'clock in the morning on 6 August the American Air Force had detonated a bomb of unheard-of power over a Japanese city. It was two thousand times more powerful than bombs of the same weight carrying conventional chemical explosives. In a city numbering around 400,000 inhabitants, hundreds of fires had broken out. The city itself was almost totally obliterated. The bomb had utilized previously unknown physical principles for converting the energy of an atomic nucleus into mechanical, light and gamma-radioactive energy.

I could barely catch my breath. And so Lev Landau had been right! When I'd met him in November 1944 he had said that the time had come to take from life everything you possibly could, because the end of the world might be coming. Theoretical physicists did not doubt the possibility of creating an atomic weapon whose power per unit of weight would be a thousand times greater than that of conventional chemical explosives. For the present it was not known whether there would be any defence against such a weapon. For the theorists, everything was now clear. All that remained were a number of engineering and technological problems. When these were solved, the world would have a superweapon at its disposal. But for some reason it seemed that the atomic bomb belonged to the remote future. In any case its creation and use would not have any bearing on the Second World War. It had been possible to destroy some hundred million people using the old means. What more was needed?

Trying not to wake Zina, I rushed over to the telephone and called Al'tshuler. 'Lev, listen. Dau[14] was right. They've made that infernal bomb. Not only made it but dropped it on some Japanese city.' Al'tshuler lived nearby, on Chistyi Lane, about ten minutes' walk from our house. Both of us had a fairly good idea of how things stood with our own army's munitions. Creating that monstrous bomb here, in a country whose economy had been wrecked by a devastating war that had only just ended, would not be easy...And nevertheless...If we could break the back of Hitler's gigantic war machine, why could we not catch up the Americans in this new field of science and technology?

At work all conversations revolved round the American 'surprise'. I rang my theoretician friends at the Institute of Physics. No, they did not think the TASS communiqué had been a canard. It followed from Einstein's equation that, during the conversion of matter to energy, forces would be released which were close to those in the Truman announcement. And yet none of us expected that a lone aircraft could drop on a defenceless target a modestly sized bomb, the explosion of which could have such appalling consequences. The reports published later about the shadows of people burned into the asphalt and the walls of houses staggered the imagination.

---

[14] Nickname for Lev Landau.

But how had the Americans managed to overcome all the difficulties of creating a superbomb? Lev Al'tshuler reminded me: 'You mustn't forget that during this war the Americans did not fight on their own territory. Their casualties were nothing compared with ours. Safe on their side of the ocean, they kept working away quietly. Or rather, not so much quietly as calmly.'

An important event of the time was the publication, in a large print-run, of the Russian translation of Smyth's *Atomic Energy for Military Purposes*. This book clarified a great deal for those of us who were professionally close to its subject matter. We understood very clearly that the success of such a complex undertaking was helped by the conditions in which American scientists worked. Their research and development attracted huge resources. This was not possible in the Soviet Union, as long as the war continued. However, whatever the circumstances, the atom bomb had become a reality. It was in foreign hands and was a threat to peace. It became our task to create our own atom bomb.

TALKS WITH KHARITON

AT the end of December Iulii Borisovich Khariton came to our laboratory and asked, without further preamble: 'Have you read Smyth's book?' 'Of course.' 'Then you realize what a huge amount of work will have to be done before our country has the secret of atomic weapons. I would like your laboratory, which has been working on the radiography of explosions and detonations, to engage itself fully with the atomic problem. Don't worry about the formal side of things. All I need is your consent.'

We asked for two or three weeks to think things over, although we did give him our tentative consent.

In January 1946 Al'tshuler and myself received a State Prize for the invention of flash radiography of gunshot and explosion phenomena. There were congratulations, embraces from our friends and telegrams. Among these one in particular drew our special attention. It was signed by Igor Vasil'evich Kurchatov. By then we already knew that Kurchatov was head of the Soviet atomic project.

At the beginning of February 1946 Khariton made another visit to our laboratory. This was a working meeting at which problems connected with the change in the laboratory's research specialization were discussed. At the end of the meeting he said: 'In order to complete the development work experiments will need to be done which will be difficult to conduct in the conditions obtaining in Moscow. You may need to be relocated to other regions for six months, or a year, at a time. But it's early days to be talking about that. First of all we must create a base for experimental research here in the capital.'

Of course, no-one thought that the period of time outside the capital would be extended to several decades...

Part 2

ON ANOTHER MERIDIAN

In this chapter we relate how we bade farewell to Moscow, traded one meridian for another, and began a new life, full of hard work, arguments and discussions, and how we travelled the whole distance from preliminary research to the first testing of an atomic bomb.

This section of the book contains a good deal of specialized information. Without it it would be impossible to convey the essence of the story of the solving of the atomic problem. Do not be put off if not all the technical details are comprehensible to you.

THE BEGINNING OF A NEW LIFE

IN May 1947 my grandmother, Irina, Zina and I went straight from the Morozov Children's Hospital to a small place which, before the Revolution, had been the location of the Sarov monastery, one of the most famous in Russia. We had arrived in what for us was a new world. Everything was unexpected: the thick forest, the beautiful, centuries-old pines, the monastery on the high river bank with its cathedrals and white bell tower. And, in sharp contrast, the grey columns of prisoners who went through the village in the morning and in the evening.[1] Local folklore abounded in stories of vast crowds of pilgrims, of miraculous cures performed by St Seraphim of Sarov, of a visit to the monastery by the last tsar.

Nowhere could have been more suitable than this place for the grandiose installation which was being built. The large tracts

---

[1] These were prisoners from the many camps in the Sarov area. Ordinary criminals or genuine political prisoners were not used as forced labour on sensitive sites such as Arzamas-16. See 'Introduction'.

of impenetrable forest and the absence of any habitations nearby created favourable conditions for carrying out explosions on the required scale.

Rides were cut into the forest, a group of Finnish cottages[2] was assembled and some two-storey wooden cottages built. Our family settled in one of them. The family of Aleksandr Dmitrievich Zakharenkov, the future deputy minister, arrived a little before us. At almost the same time the families of Diodor Mikhailovich Tarasov and Lev Al'tshuler arrived.

The design shop, administration and dining room were housed in an old brick building. A small machine factory was already operating here, with a forge, foundry, and supply of tools. During the war it had made shells for 'Katiusha' rocket-launchers and other munitions. Our laboratories took over one wing of this factory.

Armoured reinforced-concrete bunkers for conducting explosives work had been built in the forest by prisoners. On this same site, two weeks before our arrival, the cladding had been taken off a large reinforced-concrete 'barrel', intended for experimental explosions using one- or two-kilogramme charges. In May of that year Tarasov carried out our first experimental explosion in that barrel.

The fantastic possibilities with which we had been furnished placed obligations on us. We felt inspired by the good will and profound interest shown by Khariton, Kurchatov, and the head of the installation, Pavel Mikhailovich Zernov. I was appointed director of the section which was concerned with devising research methodology for explosive processes and, above all, with developing flash-radiographic techniques. In this initial period we had to find solutions for the most varied questions which had a bearing on the whole project. Thus we had to solve the difficult problem of devising safe and reliable electrodetonators. These were immediately widely used in all explosive experiments. We were also concerned with the planning of facilities for carrying out explosions and in selecting personnel.

---

[2] A form of cheap, prefabricated, quickly constructed housing, built out of wood and without foundation ditches. The ones at Arzamas-16 were apparently obtained as war reparations after the Russo-Finnish war of 1940. In Russian the word *kottedzh*, stressed on the second syllable, denotes something considerably larger than its English equivalent 'cottage'.

In Moscow Al'tshuler and I had worked in the same laboratory. But here at Arzamas-16 it became clear that the scope of the research and development which lay before us was so great that it made more sense to assign to Al'tshuler an independent section with his own staff. That division used our shock-wave techniques for studying the properties of substances under high and superhigh pressures and temperatures. Models of various construction designs were tested there. This team of scientists working in close co-operation often brought forward new ideas and developed them. The author of many of these ideas was Iakov Borisovich Zel'dovich. Al'tshuler and I felt his mobilizing influence in full measure.

Khariton used to say: 'We have to know five, ten times more than we know today. That is the only way a scientific project can be begun and quickly perfected.'

The installation where we would be working, the future All-Union Research Institute for Experimental Physics (known by its Russian initials as VNIIEF), now called the Russian Federation Nuclear Centre, was at the very hub of things. In order to learn how to obtain at the critical moment flashes 'brighter than a thousand suns',[3] we had to solve a multitude of scientific and technical problems. First of all, our theoreticians and experimenters had to determine the pressures which would arise during the detonation of powerful explosives. The explosion products, which formed during this process inside the constructions we built, played the same role of 'working body' that steam plays in turbines and other heat engines. It was no less important to determine the density of metals when subjected to pressure surges capable of bringing fissile elements to supercritical states in millionths of a second. As a result of this research, a new scientific discipline arose—the physics of high-energy densities.

To carry out measurements of surge pressures in explosive substances and metals we proposed and developed three basic methodologies: flash X-ray exposures; an electrical contact method—the sequential closure of electrical contacts by the action of either shock waves or moving bodies—and superfast photographic recording.

---

[3]The phrase is the title of Jungk's book (see 'Bibliography'). A Russian translation of this appeared in 1961. The phrase originated in the Hindu *Bhagavad-Gita.*

We obtained the clearest and most direct information from flash radiography. The X-rays we produced delineated the displacement of gauges entrained by explosive products and allowed us to 'see' the dimensions of metallic cores at the moment of maximum compression by pressures measuring in the millions of atmospheres.

In order to begin our work on radiographic methodology as soon as possible, in March 1946, in Moscow, we attempted to place an industrial order for a 500-kV flash-radiographic system. We drew up the technical specifications and sent it to organizations doing similar work. We got refusals all round. It was pointless to waste time on negotiations. There was only one solution—to make the thing ourselves. The basic element in such a unit consists of high-voltage condensers. We hunted for them everywhere. At the 'Condenser' factory in Serpukhov we found three capacitors with the same specifications as those donated by Kurchatov in 1942. We reached an agreement with the factory management, loaded the condensers into our little MK company car, and, holding on to one of the rear doors that would not close, we drove through the deserted outskirts of early-morning Moscow and delivered the condensers safely to our institute.

Our source of X-ray flashes in the unit was, as before, a kenotron. If you temporarily increase the filament voltage to such a rectifying tube, it turns into an intense source of X-ray flashes. We finished the assembly and testing of the unit in early 1947. In May of the same year the unit was flown to Arzamas-16. Immediately we had to come up with instruments capable of recording phenomena lasting millionths of a second and less. These were high-speed oscillographs and photochronographs. We custom-built these instruments as well.

Soon after our arrival we had to take part in an emergency repair operation to ensure the energy supply to our work sites in the forest. At the time our energy source was an American turbo-generator acquired through lend-lease. During shipment via Vladivostok, the generator windings had been damaged by exposure to seawater. This led to frequent breakdowns. The power supply not only to our living quarters but also to our work facilities would be interrupted. Together with Arkadii Adamovich Brish, I had to mend the American generator. The frequency of the alternating current in that small power station would 'wander' within a relatively wide range. At Khariton's request we set up a reed-relay frequency-indicating meter with a voltameter in his office. He could now regulate the power station directly from his office.

In 1947 a number of young scientists joined our department: Brish, T.V. Zakharova, K.K. Krupnikov, S.B. Kormer, I.Sh. Model, and M.A. Manakova. V.V. Sof'ina was somewhat older. These were people engrossed in the task before them, full of energy and enthusiasm and they formed the core of our group. Relations were warm and friendly; at times it seemed we were one big family. Three graduates of the Moscow Bauman Higher Technical School joined Al'tshuler's division: Boris Ledenev, Ania Bakanova, and Militsa Brazhnik, as well as Diodor Tarasov and Al'tshuler's wife Maria Parfen'evna Speranskaia. For the young women on the staff, isolated from their friends and relatives and making their way in the world for the first time, Maria Speranskaia's house became their home. She herself was one of our leading explosives experts. One can judge the nature of her relations with family, friends and colleagues by the inscription on her gravestone: 'To the woman who gave us her heart.'

We had been panic-stricken when this large influx of 'rank-and-file, untrained' specialists suddenly came into our division. I even had words about it with our sector chief, Kirill Ivanovich Shchelkin: 'In Moscow I was working with five, or, at the most, ten people, and now I have more than twenty.'

Shchelkin tried to reassure me: 'In Moscow I, too, only had a few people in the laboratory, but now look at them all.'

In fact, we coped. Twice a week, once on a Monday and once on a weekday, we held seminars for our 'rank-and-file, untrained' staff at our house. These sessions were faithfully attended by all the staff members of our divisions. I used to give talks on explosions and detonation and tried to introduce the staff to microseconds and cosmic velocities. The situation was eased considerably by the existence of our core of 'old hands'. Tat'iana Vasil'evna Zakharova already had a great deal of experience of experiments with explosives, dating from the war years. She had done a great deal of work with as experienced an explosives expert as Khariton. Samuil Borisovich Kormer had graduated from the explosives department of the Artillery Academy.

The summer of 1947 was a hot one, both literally and figuratively. Our scientific divisions quickly grew and came up to full strength. A radiochemistry division headed by Al'fred Ianovich Apin and Vitalii Aleksandrovich Aleksandrovich became our neighbour. My acquaintance with them had begun in Moscow in 1946.

In the spring of 1948 the scope of our activities widened considerably. A theoretical section was opened, headed by Zel'dovich. Other members of it were: D.A. Frank-Kamenetskii, V.Iu. Gavrilov, G.M. Gandel'man, and E.I. Zababakhin. N.A. Dmitriev arrived in the autumn. From 1950 onwards I.E. Tamm and Andrei Sakharov worked in this division.

The laboratory premises were crowded. The heads did not have separate rooms and sometimes the staff did not have separate desks. But the work was enjoyable and very productive.

From 1947 to 1949 various units and instruments were built and put into operation. Flash radiography continued to be one of the prime methods used. There was an increase in the energy of the radiographic quanta we employed and our means of recording an 'instantaneous' image in X-rays were perfected. In 1948 we proposed and built our first sharp-focusing flash tube with an anode in the form of a needle. This significantly increased the crispness of the image and enabled us to obtain up to eight consecutive shots of the development of the same process. Kenotrons became obsolete as sources of X-ray flashes. A needle-shaped anode hardly differing from our own is still being used in flash-radiographic tubes, both in our country and abroad. Using flash radiography, we succeeded in recording a two-fold compression of iron. This meant that the iron's specific gravity increased from eight to sixteen grammes per cubic centimetre. Record compression levels were registered for copper, aluminium, and other construction and specialized materials. It was less than three years since Al'tshuler and I had rather prematurely announced at a conference in Moscow that no forces existed in Nature capable of compressing metals to such a degree. In the beginning we didn't have faith in such miracles.

An important step in the development of radiographic technology was the installation, carried out in record time, of the most powerful flash unit of its day, harnessing two thousand kilovolts. Three months after beginning installation work, we were already able to begin radiographing model charges.

An active participant in this work was Veniamin Vol'fovich Tatarskii, the head of the radiographic laboratory of a Moscow institute. A modest and charming man, he very quickly won the love and respect of all of us. Unfortunately, his life was brief—he suffered from a serious heart complaint.

Among the pioneers of flash radiography one should mention Maria Alekseevna Manakova. During the first months of the division's existence the range of her duties was extremely wide. It was Manakova who, along with Tat'iana Zakharova, made the first radiographic studies of various model charges. As I write these lines, I can picture a young woman with a chain of electro-detonators hung about her neck like a necklace. At the time this seemed to be the safest way of handling these devices for triggering explosions. I remember one particular episode: Khariton came to the bunker. To make a graphic record of the dimensions of a particular charge, it was decided to take a photograph of Maria Manakova standing next to the charge.

'Here, let me take it,' Khariton offered.

The photograph turned out superbly. Documents and the photo of the charge, with Maria alongside it, were sent via the appropriate channels to Moscow. At some point the photo came to the attention of Lavrentii Pavlovich Beria. He considered that the identity of a researcher involved in our work would thus be revealed and took steps to remove Maria from the photo and to destroy the negative.

All our work was carried out in close collaboration with Al'tshuler's division. This was useful for both groups. Our priming circuits and methodological proposals quickly became the property of his division. On the other hand our joint discussions and the profound theoretical and experimental work done by Al'tshuler and his group provided us with a sound physical orientation, and in their turn stimulated new proposals and developments.

In mid-1948 there was an 'exchange' of researchers: Krupnikov and Kormer went over to Al'tshuler's division and Tarasov came over to ours. Over the next twenty-five years he would successfully direct all our radiographic research.

This same period saw the organization of the production of many ancillary laboratory instruments and materials in our country. Good oil-vapour diffusion pumps appeared, as did counters—instruments for registering the intensity of radioactive emissions. We were especially glad of certain laboratory 'bits and pieces'—vacuum rubber, Ramsay's oil and other such things. As late as 1947 Soviet physicists were cutting out flat rubber gaskets for their vacuum instruments out of car inner-tubes. Getting hold of vacuum hoses was a problem. For a small section of Leibold red vacuum tubing people would give anything—from liquor, which was in short supply, to precision galvanometers. As a rule we made our own Ramsay's oil, out of rubber, wax, and vaseline. Now we got all these essential instruments and materials via normal channels, through the supplies departments.

A stream of instruments and measuring devices flowed into our divisions. Before, in the little radiographic laboratory at the Institute of Machine Science, we not only knew how to turn on any given unit or instrument, we knew all the 'quirks' of each piece of equipment and could sort out any breakdown. Now we had to master new apparatus, but we enjoyed doing so. How pleasant it was, for example, to know that, in order to measure high vacuums, we did not have to lift the vessel of a MacLeod manometer with two or three kilogrammes of mercury in it. How often we had smashed the flask with the mercury in it! The fiendish metal would get into every nook and cranny. We spent hours crawling about the floor, collecting those recalcitrant little spheres onto pieces of paper, all the while breathing in mercury fumes. Now we could get a pressure reading to a hundred-thousandth or millionth part of a column of mercury directly from the scale on the vacuum-meter. We were called on to work with our hands less and less. Our responsibility for our choice of research avenues and for their results was growing.

1948 was the year we began miniaturizing flash-radiographic generators. We built, and successfully tested a laboratory prototype of such a generator which weighed only twenty kilogrammes and used five hundred kilovolts. At that time the usual low-voltage generator weighed more than a hundred kilogrammes.

Another avenue we pursued in the miniaturization of flash-radiographic apparatus was the creation of units using relatively low voltage: sixty to a hundred kilovolts. Nikolai Vasil'evich Belkin took an active part in this work. With his help, and that of a number of other researchers in the laboratory, we succeeded in making a radiographic device weighing about one kilogramme. Konstantin Fedorovich Zelenskii, who co-operated with me in the initial stages of this research, was the first to report, at the Moscow Medical Radiological Institute on Solianka Street, on the creation of the smallest portable radiographic unit.

I recall the reaction of one elderly radiologist. 'In my whole life I have been really surprised twice. The first time was in Moscow in 1910 when I saw with my own eyes Blériot's monoplane take off from Khodynka Field. And today when we were shown a radiographic unit weighing slightly more than two hundred grammes, powered by two torch batteries.'

I had had occasion to hear kind words said about my work but this commendation for some reason was particularly pleasing.

On 15 August 1948 two events happened simultaneously: the commissioning of the new oscillograph ETAR and of the first demountable, constantly pumped flash-radiographic tube. The ETAR was built by two highly qualified radio engineers, E.A. Etingof and M.S. Tarasov. The name for the new oscillograph was a combination of the initial letters of their surnames. Electron-beam tubes produced by the German firm AEG had been found by Khariton in a warehouse of captured equipment. Oscillographs with such tubes were used successfully for many years in our laboratories. Externally the ETAR looked more like a sewing machine than a modern oscillograph. Nevertheless, it made it possible to register time intervals in the hundredths of microseconds (10^{-8} sec.)

In mid-1949 the Institute of Chemical Physics of the Academy of Sciences developed the OK-17 oscillographs, which could be used to solve our problems. The development of Soviet oscillography is closely linked with the name of one researcher at that institute, I.A. Sokolik. Through a piece of ineptness on the part of a doctor, the life of this talented designer and inventor was cut short. He died in 1960, at the height of his talent and creative powers.

Our first photochronograph was built at the very end of 1946. It was a chamber with a revolving disc and a high-RPM motor from a Maiak domestic vacuum cleaner. We got the vacuum cleaner from a second-hand shop in Moscow. The peripheral velocity of the disc on which the photographic film was mounted was not great and did not allow us to capture the processes we needed to in any detail. However, in 1948 I. Sh. Model managed to construct a photochronograph with a peripheral velocity almost a hundred times greater. We were now able to observe phenomena which occurred in ten-millionths of a second.

We did not receive any production-line photochronographs until the second half of 1948.

The first explosive charge with photochronographic recording was a failure: the synchronization of the rotating mirror did not function properly. In such cases no image is recorded on the film. Everyone was bitterly disappointed, especially the director of the experiment, Vladimir Stepanovich Komel'kov. He set out on foot on the ten-kilometre walk from the forest test site to the main work site. During the two hours this took, Brish conjectured that the reason for the failure was a wrong choice of polarity. The experiment was repeated. This time everything worked well and a crisp image of the explosion was recorded on the film. Afterwards Komel'kov used to say that when he learned that the cause of the failure was so simple, he could have kissed Brish. But in those days, kisses on the occasion of experimental successes were not yet the done thing.

THE MIRROR

IN September 1947 we obtained good photochronographs of the explosion of comparatively large charges. However, other difficulties lay in store for us in this area. As a rule, shrapnel from the charge casing would go through the embrasure and into the entrance lens of the chronograph, shattering it. Wasting an expensive lens on every experiment seemed an inadmissible extravagance. We knew what we had to do: turn the charge through ninety degrees and set a flat mirror at an angle of forty-five degrees to the axis of the charge. In this way only an ordinary mirror, some ten times cheaper than fast long-focus lenses, would be broken. But we had no mirrors with the right dimensions.

Chance came to our aid. In the town a barber's shop had recently opened. The director of the institute, Major-General Pavel Mikhailovich Zernov, made sure that his staff were always clean shaven and well trimmed.

Once, I went into the barber's and discovered that, apart from the two mirrors with which our guardians of masculine beauty were equipped, there was another large mirror, of uncertain purpose, hanging by the door. 'Mikhail Ionovich,' I said to the manager of the shop, 'please lend me that mirror for one night.'[4]

Mikhail Ionovich sensed something ominous in this request and flatly refused. I headed straight for Zernov's office. He used to see staff immediately and at practically any time of day. After he had listened to my request, he simply asked: 'And when are you thinking of returning it?' 'Never. We're going to destroy it tonight. But I've already placed an order with the supplies

---

[4]The man's surname was Fedorov.

department. We need dozens of mirrors for our work.' Zernov thought it over for a few seconds, then said: 'All right, let's go to the barber's shop and have a look at this mirror without which your work can't progress.' A few minutes later Zernov and I were with Mikhail Ionovich. Seeing me in the company of the general, he launched a counterattack. 'Pavel Mikhailovich, this is daylight robbery. We only got our mirrors last week. The salon's just begun to look respectable and they're taking them away.' But PMZ (that was our abbreviated name for Zernov at the time) was adamant. 'Give the mirror to V.eniamin. You'll get a new one from Moscow.' Kormer and I immediately loaded the precious mirror into the car.

I need hardly add that after this episode I was *persona non grata* in the barber's shop. I had to shave with an old safety razor. It took about a year for diplomatic relations with Mikhail Ionovich to be restored.

SAFETY PROCEDURES

THE majority of the physicists and engineers in the experimental divisions had never worked with explosives. The courses I arranged in my own home, when I trained future specialists in the characteristics of various explosive substances and in the safety procedures for working with them, partly eliminated this gap in their knowledge. At that time our favourite explosives were lead azide and fulminate of mercury. Vapours from lead and mercury— elements at the end of Mendeleev's periodic table severely weaken X-ray radiation and result in contrast shadows on radiographs.

Experiments involving explosives always call for special care and attention. To this day I am still surprised that in our division there were practically no accidents. Naturally, instructions had to be carried out to the letter. At work sites, alongside these instructions were posted notices: 'Explosives staff, remember, you're not entitled to make mistakes' and 'Pay attention. When the Lord God created Man, he did not supply spare parts'.

There are also numerous unwritten rules which enhanced the safety of explosives work. Aleksandr Fedorovich Beliaev told us about some of them. They have stuck in my mind ever since. For example, here is how he recommended working with detonating cord during preparation for an experiment. In this sort of work the dangerous moment comes when you cut the detonating cord to the correct length. The laboratory assistant or demonstrator should not have their legs beneath the work bench where the work is being done because there is a chance that the coil of cord which remains under their feet will explode.

Unfortunately we learned too late that the priming devices we were using were very sensitive to electromagnetic induction and to electrostatic charges which can develop under friction. This led to a number of injuries sustained during work with electrodetonators.

During this first stage of our work not much attention was given to safety procedures. A charge would be hung in front of the armour-plated bunker in an ordinary string shopping-bag. A few preliminary X-rays would be taken in order to confirm that the charge was orientated in line with the X-ray beam. Then Maria Manakova would come out of the bunker and bang a hammer against a length of rail hung from the branch of a tree. The rail had been left by the builders. These signals meant that an explosion was imminent and everyone out in the open had to take cover. Sirens, telephones, and other 'miracle' warning and communication systems appeared only later. Communication with the work sites was achieved basically through the security guards' field telephones. There was no direct link with the bunkers. There were many curious incidents. For example, we heard the words: 'This is Sergeant Kurochkin. Your workers left their nappies behind and would like someone to bring them.' We had to work out that this meant that our researchers ('workers') had forgotten their photographic film ('nappies'), without which they could not conduct a single experiment.[5]

In January 1948 Georgii Pavlovich Lominskii arrived. He had been appointed head of the explosives sites. This was an extremely timely appointment. Before Lominskii, the heads of these sites (among them the author of these lines) has been chosen at random. A specialist in explosives and priming devices, Lominskii, was well respected and well liked. He always understood that, apart from safety procedures, which he always rigorously observed, there was a common cause, for which we were all working. He tried to look at every complex experiment not in a formal manner, but in a human manner.

---

[5] The sergeant has confused two Russian words: *plenki* (films) and *pelenki* (nappies).

Gradually, the organization of explosives research began to fall into place. A dispatcher service was introduced to the work sites, as were storage facilities, small buildings for preparing explosive materials, and a thoroughgoing alarm system. This work was no longer considered exotic—experiments with explosives were becoming routine.

Experimental explosions on the work sites in the forest went on day and night. Several groups were working simultaneously.

All the same, recalling those days through the prism of the intervening decades, I have to admit candidly: we fell on our feet. It was only by sheer chance that many experiments did not lead to serious injury. Lady Luck often saved our experimenters from the consequences of uncontrollable explosions.

The first such 'unmanageable' explosion occurred in Kazan during the war. On 8 March 1943 Zina and I were setting up a small charge of about three grammes in a covered passage. All that remained to be done was to connect the leads from the detonator capsule to the blasting cable. Zina was getting ready to do this. 'Wait a second while I check the high tension lead again.' I had not had time to switch on the voltage and bring it to the required magnitude when the whole arrangement blew up, with the detonator capsule still unconnected. By a miracle Zina escaped injury. At this early stage of our work we did not know that electrical inductions can set off even an unconnected detonator capsule.

A second incident took place at the beginning of 1948. We were preparing two experiments simultaneously. Boris Ledenev and Ania Bakanova were setting up a charge of about two kilogrammes in the reinforced concrete barrel. At the same time in the neighbouring facility I was performing test switching of a flash-radiographic apparatus. Suddenly there was a powerful explosion. Everyone in the shelter realized what had happened: the charge on which Ania and Boris had been working had blown up. Our hearts stopped. A few seconds passed, which seemed like an eternity—and in the doorway of the shelter appeared the agitated Ania and the imperturbable Boris. 'Nothing out of the ordinary,' he said. 'That was our charge going off because of your induction. We'd already moved away from the barrel.' I sank into my chair, stunned. The thought went through my head: 'You know very well that, when you switch on high-voltage equipment, the levels of induction that develop along the cable trunkings are enough to set off an electrodetonator.'

After this incident, in all lists of instructions a clause appeared prohibiting any kind of work with high-voltage equipment during the preparation and conducting of experiments involving explosives.

In those distant days we often had to work with charges of different composition and form. Usually these were mixtures of TNT with some powerful secondary explosive. In the ventilation cabinet of one of the laboratories in Al'tshuler's division a water bath had been set up, in which the TNT could be heated up to its melting point, after which the powder of the secondary explosive would be added to it. We knew that the temperature of the melted TNT should never exceed 90 degrees. Any hotter and it can ignite. That day the monitoring of the bath temperature was being done by two young female graduates of Moscow's Bauman Higher Technical School. The young women failed to notice the temperature, and the molten TNT caught fire. Everyone in the room panicked and made a dash for the exit. Diodor Tarasov was the only one not to lose his presence of mind. Acting strictly according to the instructions, he quickly poured the burning TNT out onto the floor, and when it had spread out into a thin layer which could not detonate, he extinguished the fire with sand. After this incident, melting TNT, making TNT-based mixtures or storing explosives was categorically forbidden on laboratory premises.

We checked especially carefully to see that no explosives were kept in laboratories in the days before a public holiday. An amusing incident occurred when we were checking the laboratories on the eve of one May Day. In one workroom a few kilogrammes of a white powdery substance were discovered. As per instructions, it was taken away to a work site in the forest to be blown up. This strange explosive substance, however, did not blow up. Further careful analysis revealed it to be wheat flour, given out to members of staff for the holiday.

A memorable incident occurred in Samuil Kormer's group. They were preparing an experiment which involved an explosive charge of over a hundred kilogrammes. Suddenly the charge caught fire. In such circumstances combustion can lead to detonation, with all the consequences which flow from that. Samuil stayed calm and collected. He led his team away to the shelter, phoned the dispatcher, and forbade anyone to approach the seat of the fire. Nature was kind: the charge burned down safely and there was no explosion. Afterwards there was much argument about the causes of the spontaneous ignition of the charge. According to the official version, the charge caught fire as a result of the focusing of rays of sunlight in a drop of liquid left by a passing bird.

FLIES

FOR work with radioactive substances, special premises were set aside. They had thick partitions which divided the room into separate compartments. It was quickly discovered, however, that they did not afford sufficient protection, and for work with highly active substances it was decided to equip special premises on the same side as the main building. To a degree the method of working in these premises was reminiscent of 'hot' radiochemical laboratories. Here is a curious story about our work in these new premises.

According to regulations, on finishing work with radioactive materials, the premises had to be handed over to the commandant of the military security unit. Usually this procedure took up a great deal of time—you had to phone the commandant's office, call out both the commandant and the guard and then wait for them to come all the way from the checkpoint to our building. During all this time the people doing the experiments would amuse themselves by catching flies, of which there was a particularly large number around the windows on sunny days.

Once, the commandant, who had arrived with the guard detail, indicated the heap of dead flies on the windowsill. 'What's this you've got here?' he asked the attendant on duty. 'How do you mean—what?' asked the attendant in his turn. 'They're flies.' 'I can see they're flies, but look, they're all dead.' 'Yes, they're dead,' the attendant confirmed. 'Well, and how about yourselves?' 'So far, we're still alive' said the attendant, who had begun to grasp what the commandant was afraid of.

This time the transfer of the building to the military was done unusually quickly, and from that day onward the commandant never appeared in our building. He delegated this task to his aides, clearly thinking that anyone who worked in the building was a dead man.

The Library

FOR scientists a good library is an essential prerequisite for fruitful work. Khariton, who understood this very well, devoted much attention to creating one. In March 1947 he invited a qualified librarian, Elena Mikhailovna Barskaia, to work in Arzamas-16. This is what she says:

...I had to begin from scratch. In building up our stocks, Khariton advised me to use my experience in organizing the library of the Academy of Sciences' Institute of Physical Problems, where I had worked during the war.

In our new workplace myself and the director of the public library Rufina Nikolaevna Alekseeva were given a room with two desks and two empty bookcases. We were soon given the right to get books and journals from the State Scientific and Technical Library and from the library of Moscow University. Foreign journals could be bought for hard currency. We also had the opportunity to choose from literature brought out of Germany as part of reparations. This material was stored in the yard of the Peter-Paul Fortress in Leningrad. There the journals and other literature were lying scattered all over the ground and I had to sort them out. I was in regular telephone contact with Khariton to get advice as to whether we needed this journal or that. As a result of our efforts the library facilities available to the staff at Arzamas-16, as far as books and journals were concerned, were the equal of anything in Moscow.

I remember one episode. In the Peter-Paul Fortress I had packed away the journals I had selected into wooden boxes, nailed down the lids and sent them off by rail. I went home. About ten days later they phoned through a telegram to say that books in foreign languages were lying about the railway tracks. They put me in an open-topped light aircraft and we flew to the 'location of the occurrence'. There I found my boxes all smashed and English, German, and French journals scattered all over the place. I collected them up and packed them again. Soon they reached their destination safely.

In order to discuss how to stock the library and how to classify the materials, we managed to organize a library committee. The members of this committee always acted both as consultants and friends. I remember with gratitude the first chairman, D.A. Frank-Kamenetskii, Ia. B. Zel'dovich, I.E. Tamm, G.N. Flerov, and V.A Tsukerman. In all our difficulties Iu. B. Khariton was, and still is, a great source of help.

Among the specialists of various disciplines Khariton selected for our work some with a good knowledge of foreign languages. Among them was Sergei Ivanovich Borisov who, before the war, had been a fifth-year student at the Institute of Foreign Languages. This is how Borisov described his first meeting with Khariton:

> I was taken through into a large study where two people were sitting at a desk. One of them was a tall, thick-set military man holding the rank of Major-General. I saluted him and addressed him according to the regulations. The general indicated with a gesture that I should introduce myself to the civilian who was sitting next to him. I had barely had time to do this when the civilian turned to me and began speaking in first-class English. This was so unexpected that I was immediately flummoxed by these English words, although I understood the gist. Khariton realized my confusion and went over to Russian. I gradually got the hang of it and was able to answer him tolerably well in English. After four years in trenches and dug-outs, I was hearing English in its purest form for the first time.

ROMANCE AND LIFE

DURING the first, most romantic days of our life at Arzamas-16, a remarkable atmosphere of good will and support was created around our research. We worked without heed for ourselves, with huge enthusiasm, mobilizing all our spiritual and physical strength. The working day for senior researchers lasted from twelve to fourteen hours. Zernov and Khariton worked even longer hours. There were practically no days off, nor was there any leave; permission to travel on business was granted comparatively rarely.

Joint seminars of theoreticians and experimenters were held regularly, with Khariton unfailingly present. The seminar topics were varied: they included nuclear physics, methods of investigating fleeting processes, special areas in gas dynamics, and issues in the creation and measurement of high and superhigh pressures. Zernov would often appear at these seminars. I remember his frequent visits to sites where experiments with explosives were being conducted. He made on-the-spot inspections of the current state of development projects and the assembly of new units. Once, when Maria Manakova and I were in a bunker, completing the shelter's protective cladding, in came Zernov, expressed his surprise at the rapid progress of our work, congratulated us and asked: 'What else do you need to get the two-million volt unit running sooner?' 'Some castor oil would help. Only we need a lot, about a hundred and fifty kilos.' Two days later Zernov's secretary called us at the laboratory. 'You've been sent a barrel of oil from Bulgaria, about two hundred kilos. You can pick it up at the warehouse.' Other supply problems were dealt with just as efficiently. Zernov stayed abreast of everything which the experimental laboratories were working on.

Work involving restricted documentation and materials called for special care and attention. Such work did not always go smoothly. One memorable incident occurred in December 1949 in Viktor Davidenko's division. A scientist, on finishing his shift, wrapped up a crucial component, about the size of a walnut, in aluminium foil and forgot to put it away in the safe. The next morning the cleaning lady took it for a sweet wrapper and swept it into the waste-paper basket. The rubbish was then taken out into the woods and buried.

All the men in Aleksandrovich's, Davidenko's and Apin's divisions went out into the freezing cold in their sheepskin coats and methodically combed through the snow where the rubbish was buried. It was three days before their efforts were crowned with success.

This happy outcome so pleased everyone that the divisional heads involved threw a banquet in the recently opened restaurant. Times were harsh—the chief culprit could have faced arrest if the part had not been found. As it was, he got away with a reprimand, signed by Khariton. This seems to have been the only occasion on which our scientific director personally signed a reprimand to one of his staff. Youth would have its day. We found time for brief relaxation. Many members of staff were still single and without families. The heads of the institute, Zernov and Khariton, were forty-four. The average age of the research staff was twenty-eight. On the rare Saturday or Sunday evenings when we were free, people would gather in the homes of those with families. We would dance, read poetry and sing:

> From the wind and the cold
> We've begun to sing worse.
> But to those who reproach us we'll say,—

Do some pumping with us,
Take some photos with us,
Blow things up with us,
Just for a year.

Some slightly altered lines from Pushkin's *Poltava* were no less popular:

The Bearded one is rich and famous,
His installations can't be numbered.
Herds of scientists wander there,
Although they're free, they're guarded.[6]

There were no decent record-players or tape recorders. We made do with ancient gramophones. They were often breaking down. I would sit down at the piano and play foxtrots, tangos and waltzes. The mahogany piano, which belonged to our family, had also been brought from Moscow. It was the first musical instrument in the town. Sometimes we discovered that several couples were dancing on the asphalt path outside to the strains of my rather primitive accompaniment.

There were competitions for the most inventive and apt toasts. I remember some of them: 'To our fair Moscow, which can live and work in peace as long as we live and work here!' 'To respect for the numeral in the course of absolute measurements!' 'To Her Majesty, Queen Science!'

Usually these parties were linked with production achievements. We kept a special tally of successful and unsuccessful experiments using borrowed sports terminology. If people came back from the test sites with a score of 2:1 in favour of Harry Truman, this meant that out of three experiments, two had been unsuccessful.

Sometimes on Sundays, depending on the time of year, we would have ski outings or would arrange picnics on the bank of the river, with bonfires, songs, and swimming. This was long before the era of universal car-ownership. However, a few

---

[6]The opening lines of the first canto of Pushkin's narrative poem *Poltava* (1828-29) read:

Kochubei is rich and famous,
His fields vast and boundless;
Herds of his horses
Graze there, free and unguarded.

members of staff had acquired motorcycles. Aleksandrovich and Zel'dovich had clubbed together to buy a powerful Harley Davidson with sidecar. The way they divided their responsibilities was rather strange: Zel'dovich did all the driving and Aleksandrovich did all the repairs. The experimenters used to tease the chief theoretician about his utilitarian attitude to the motorcycle. He would smile and say: 'The machine has to go. Nothing else matters.' Many was the time I returned late at night from the factory site in the roomy sidecar of that motorcycle.

Thinking back on those times, I can picture the following scene: it is a clear Sunday morning. A large number of experimenters, wearing trunks or bathing costumes, are cheerfully throwing a volleyball to one another on the river bank. Zel'dovich pulls up on the Harley Davison with a jaunty tailspin. Samuil Kormer asks him: 'Let me have a ride on the luggage-rack, please.' 'Fine, hop on', Zel'dovich proposes amiably. He opens the throttle and carts the half-naked Kormer right across town to the Red Hotel where the unmarried men lived at this time.

There was a great vogue for practical jokes of various kinds. The theoreticians were especially famous for them. First prize for inventiveness undoubtedly went to Zel'dovich. For example, a single galosh was cunningly fixed over the entrance doorway. A system of strings was arranged in such a way that when the door opened the galosh fell on the head of the person who was coming in. In Al'tshuler's office there was a placard borrowed from the prisoners: 'Remember this instructive rhyme—work enough, you'll do less time'.

Humour, sometimes extremely gloomy and crude, helped to alleviate excessive stress. On 10 June 1953 the newspapers and the radio carried a brief announcement about the arrest of Beria. It so happened that one of Zernov's deputies, the director of the explosives factory Anatolii Iakovlevich Mal'skii, with whom Khariton had worked since the war years, was the first to hear this news. In the middle of the day Mal'skii stepped into the office of V.I. Detnev, our institute's plenipotentiary of the Council of Ministers. Detnev was sitting under a large portrait of Beria and had no inkling of the latest events. 'Vasilii Ivanovich, what are you doing sitting under that scum?' Mal'skii asked him. The effect of this question exceeded all expectations. Mal'skii, who had a well developed sense of humour, told us later: 'Detnev leapt out of his armchair, his face contorted, his eyes popping out of his head, and spluttered: "Are you out of your mind, or what?"' It was a splendid spectacle.

Cultural life was gradually improving. Films began coming in. At first they were shown in a corridor of the hotel. Soon the Moskva cinema opened. During the early years this building was also used for ceremonial gatherings in honour of revolutionary holidays. In May 1949 our drama theatre opened.

The romance of our work constantly impinged on the lives of those who lived around us. Sometimes the most unexpected problems arose, and had to be solved without delay.

The following astonishing incident occurred much later, in 1978. But confidence in our eventual success, and the high degree of activity which characterized the first stage of our work, had stayed with us and helped to save a young woman from certain death.

At first the diagnosis did not appear tragic. Liudmila Golubeva had had attacks of bronchial asthma before. She had been suffering from this treacherous disease for seven years. On 21 February 1978, when Liudmila was admitted to the hospital's intensive care unit, the doctors described her condition as 'moderately serious'. But hormone treatment, and other medicines prescribed in such cases, proved ineffective. The ominous symptoms of her disease, difficulty in breathing and bronchitis, were not abating but growing worse with every passing day. Her temperature rose to thirty-nine degrees. After forty-eight hours she was put on a respirator, but even that did not bring about the expected improvement in her condition. In spite of supplementary massage, air could no longer be heard entering the lower lobes of her lungs. Subsequently it stopped entering the middle lobes of her lungs as well.

By the morning of 26 February her condition had become still more grave. She was now unconscious. The electroencephalograph—an instrument registering electrical activity in the brain—was tracing a straight line, interrupted only by occasional small peaks. The oxygen content in her blood haemoglobin fell to a critically low level, four times lower than normal. Acute oxygen starvation—the medical term is hypoxic coma—was bringing a fatal outcome ever closer hour by hour. Only a faint reaction of her pupils to light showed that there was a glimmer of life left in her. Death was imminent. Her parents were summoned to the hospital to take their leave of their daughter.

Everything that followed could, with justification, be called a miracle. At nine in the morning on 26 February I rang the intensive care unit. The chief doctor there, Anatolii Borisovich Semin, took the call and told me that Liudmila could possibly be saved only if they could get her into pure oxygen or else oxygen-enriched air at an increased pressure. But the hospital had no such pressure chambers. We decided to use our own resources to try to design and construct such a pressure chamber in as short a time as possible. First of all we dispatched a large polyethylene sack to the hospital. They placed the patient in the sack and filled it with pure oxygen to a pressure of one atmosphere.

Meanwhile the search was already on for the necessary components to make a hyperbaric chamber. Suitable offcuts of piping turned up among Kormer's supplies. In twelve hours we managed to build a chamber sixty-three centimetres in diameter and two metres in length. On 27 February the chamber and its auxiliary equipment were delivered to the hospital's intensive care unit.

We were well aware of the risks and complexities of work with such chambers. To prevent the objects placed in pure oxygen from igniting, we decided not to have any electrical leads enter the chamber itself. In order to observe the patient we made two fifteen-centimetre portholes in the faces of the chamber. The first treatment session with the chamber took place on the night of 27 February. The blood oxygen count measured immediately after the session was 1·4 times greater than normal.

On 28 February we had to win a small battle on account of having used an uncertified chamber. My own declaration that I was qualified to certify high-pressure vessels of up to two hundred and fifty atmospheres did not cut much ice with the medical people. 'We have to act in accordance with the instructions of the Ministry of Health,' announced the administration of the medical department. However, when it came to a choice between observing official instructions or saving a human life, common sense prevailed. Here, Dr Nikolai Baldin, deputy to the chief doctor in charge of treatment, was a great help. Together with Semin, he took personal responsibility for any possible consequences.

Rescuing someone from the jaws of death turned out to be a long and difficult process. It was only after several sessions of treatment that alpha- and beta-rhythms showed up on the electro-encephalogram, indicating activity in the cerebral cortex. The patient began to make very simple gestures in response to verbal prompting. Then she began to write her replies in a clumsy hand.

Our main anxiety was over: her cortex was functioning normally. It was still a long time before her swallowing reflexes and her speech returned. Two weeks passed before it became possible to feed Liudmila through a nasal tube. Only on the night of 12 March did she begin speaking in a whisper. Consciousness was completely restored. Her life had been saved.

The Dispute

THE leading laboratories at Arzamas-16 spent two years using various methods to measure the detonation pressure of explosives, on which depends their efficiency. Knowledge of this factor allowed the power of the first Soviet atom bomb to be accurately predicted and, in essence, the first test to be carried out. Theory had not provided an unequivocal answer to this question. The experimenters had to decide for themselves who was right: the German scientists or L.D. Landau and K.P. Staniukovich. The discrepancy in estimates for basic explosives was very large: 200,000 versus 250,000 atmospheres. The top priority of the experimenters was to eliminate this unknown.

V.V. Sof'ina and I obtained the first results through flash radiography of exploding charges. Instantaneous radiographs registered the path of the detonation, and the displacement of opaque tracers along the charge axis. This is how we measured the velocity of explosion products and this allowed us to calculate the detonation pressure.

In our first experiments using small charges, the tracers planted in the charges—steel balls one millimetre in diameter—remained practically motionless. However, the conclusion that the explosion products had been just as motionless, was, of course, very premature. When it was discussed it evoked a very fierce reaction from visiting Academician N.N. Semenov. He declared: 'If your methodology is not registering any velocities for explosion products, all that tells us is that your methodology isn't worth a damn.' However, the methodology was, in principle, sound. All we had to do was to increase the charge and replace the steel balls with bits of thin foil. With these 'zebra-striped' charges we obtained values for massive velocities which approximated to the estimates of Landau and Staniukovich. At present, the use of radiography of zebra charges in the study of the dynamics of explosives is widely practised by scientists in the USA, China, and the Soviet Union.

By the spring of 1948 even more direct confirmation of the predictions of Landau and Staniukovich had been obtained in Al'tshuler's division. Two independent methods had now demonstrated that the explosion-product velocity in the explosive was two thousand metres per second, and the pressure was two hundred and fifty thousand atmospheres.

Unexpectedly, in September 1948, at a meeting chaired by Kurchatov during one of his visits to Arzamas-16, it was announced that E.K. Zavoiskii's laboratory had obtained much lower values for explosion-product velocity—approximately 1,600 metres per second. That meant it would be impossible to meet the government deadline for completion of our assignment. Consequently, all our senior figures including Vannikov were very much concerned with this problem. Zavoiskii was a major radio physicist and the electromagnetic method he was proposing was founded on undisputed physical laws. Since the methods Al'tshuler and I were applying in our respective divisions were also fundamentally sound, attempts by the commission set up for this purpose to reach agreement were inconclusive.

In order to elucidate the reasons behind the low values for explosion-product velocity, Zavoiskii's methodology of electromagnetic registration was replicated in our division. In particular the material used in the tracers was changed from relatively heavy metal (copper) to a lighter metal (aluminium). After some time-consuming efforts, a series of inconsistencies in the set-up of Zavoiskii's experiments was discovered. After careful checking we obtained convincing figures for the true velocity of explosion-products (two thousand metres per second).

The heated debate which was thus concluded had not always been conducted in appropriate fashion. Zavoiskii's colleagues accused us, for instance, of treating radiographic experiments in a way which was incompatible with what they called 'materialist dialectics'. We (Al'tshuler, Zel'dovich and myself) fought our corner, but our arguments never involved ideology.

On the day of one memorable meeting Komel'kov asked Khariton what the upshot of the discussion had been. '15—love Tsukerman,' was Khariton's unhesitating reply.

By the end of the 1940s three methods for determining detonation parameters had been perfected. They subsequently formed the basis for all similar research both at Arzamas-16 and at the Academy of Sciences Institute of Chemical Physics.

CRITICAL STAGES

THE creation of new methodologies to research fleeting processes and the measurement of detonation pressure were very important tasks, but they were not the only ones we faced.

In 1948 I proposed and, together with Zel'dovich, implemented, a fundamental improvement in the construction of atomic charges. From 1949 onwards Arkadii Adamovich Brish was in charge of this new project.

One of the important features of atomic charges is the neutron source which guarantees the appearance of a neutron current in the charge after the plutonium has reached a critical state. Originally such sources were arranged in the middle of the atomic charge. We proposed an external impulse neutron source which would initiate a nuclear explosion at the moment of maximum compression. Through this we achieved an increase in power and a saving of nuclear fuel, which was in very short supply at that time.

Successful test firings of atomic charges with an external neutron source were conducted in 1954. Analysis of American publications shows that work on the creation of an external neutron source was begun at roughly the same time. However, we achieved practical application of this rather earlier than the Americans.

Among the fundamental pieces of work done between 1947 and 1950, brief mention should be made of the discovery of high electrical conductivity in explosion products and in shock-compressed dielectrics. This discovery was made by Brish, Tarasov, and myself. Before we began work in this area, the majority of researchers thought that the electrical resistance of solid dielectrics and explosion products stayed practically constant through shock and detonation waves. However, a series of specially designed experiments demonstrated that this is not the case. Under powerful compression the electrical resistance of dielectrics decreases significantly, and they become electrical conductors. At first no-one believed our results. Zel'dovich even staked several bottles of champagne on the issue. It took a variety of methods and dozens of experiments to convince not only the experimenters, but also the theoreticians, of the existence of high conductivity in dielectrics and gases subjected to powerful shock waves. The bet was lost. Zel'dovich acknowledged his mistake and jokingly suggested that the new phenomenon be named 'the Brish effect'.

Over the course of several months, each time they visited our laboratory, Frank-Kamenetskii and Zel'dovich would end every conversation with words to the following effect: 'It would be a good idea to develop a method for measuring the temperature at the shock-wave front.' This propagandizing had its effect. Model and I proposed a methodology and began taking systematic measurements of gas temperatures during explosions. These pioneering efforts are cited to this day in the literature.

One of our most pressing problems was the study of metal compressibility under superhigh pressures. To that end Al'tshuler's group for the first time applied dynamic methods based on ratios which have been known since the end of the last century. The first results on the compressibility of metals were obtained at the very end of 1947 by Tarasov. In 1948 Al'tshuler and K.K. Krupnikov managed to study the properties of a number of metals under pressures of five million atmospheres, which turned out to be the ceiling level for American researchers. In 1952 compressibility measurements were carried out at ten million atmospheres. As one American journal wrote in 1988, the means of obtaining such pressures have never been described, and the results obtained by Soviet scientists have never been surpassed.

A Raw Recruit

I RECALL one interesting incident which occurred during tests at the outer testing ground. There were four civilians there—A.A. Shorokh, A.P. Zykov, Iu.V. Mirokhin, and myself—and one soldier, our driver Lev Zinchenko. As often happens in such cases, we were running two to three hours behind in our preparations for an crucial test. It was long past midnight when we finished our preparations. The only place we could get a bite to eat at that hour was in the officers' mess. We drove there and found no-one, except for a colonel sitting in the far corner. No sooner had the waiter taken our order than the colonel leapt to his feet, came over to our table, and said to Zinchenko: 'Stand up!

About turn! Quick march!' Poor Zinchenko stamped his way to the exit. I felt my face flush red.

'How can you, an officer of the Soviet army, permit yourself to give orders like that? This man has been working with us. Has he not got the right to eat? Your immoral orders may be all right in the southern United States, but not in the Soviet Union!'

The colonel made no reply and returned to his seat. At eight o'clock the next morning I got a phone call: 'This is Kurchatov. My dear old friend, a complaint has been made against you. It says you're in breach of military regulations. What will you say in your defence?' 'Igor Vasil'evich, that colonel, whose name is not known to me, was not observing regulations but was violating the most elementary rules of human communication.' 'You know the proverb: when in Rome...' 'First of all, Igor Vasil'evich, we weren't in Rome but here in the Soviet Union. Secondly, you yourself know that I've had a white card exempting me from military service because of my eyes since before the war.' 'Now there's an idea. I'll tell the military command that asked to have you disciplined that you're a raw recruit and don't know the current regulations. I'll tell them I've had a session with you to explain them.'

The First Test

FOLLOWING the successful activation of a small atomic reactor in Moscow's Laboratory No. 2 in December 1946, in July 1948 a powerful reactor was activated at the combine in order to produce plutonium. Its production grew quickly. By mid-1949 we were already in a position to begin experiments with critical assemblies.

In early August 1949 all preparations for the first test were practically completed. Loading of the hardware, instruments and test apparatus into freight waggons began. The train set off for the Far East, following a special schedule. We travelled quickly, stopping only at major stations to change locomotives and check the rolling stock. At these stops we were surprised by the completely deserted platforms. At one station Zel'dovich, Model and a number of young members of staff, having ascertained from the conductor that the stop would last fifteen minutes, ran out onto the platform to throw a volleyball about. Zaveniagin sent a colonel out to them to order them to stop the game immediately: 'They're supposed to be serious people,' he grumbled. 'They're on a responsible mission and they behave like a bunch of eighteen-year-old kids.'

There we were at our destination. The locomotive slowly pulled the train into a zone between two rows of barbed wire. Our passes were quickly processed. We drove off in our *Gaziks* to inspect the facilities of the country's first atomic testing ground. The scale of the tests was awesome. In casemates and armoured shelters at various distances from the epicentre of the explosion, hundreds of measuring devices of every conceivable kind had been set up. The test tower, more than twelve storeys high, was an interesting edifice. The lift could raise the car carrying the bomb to a point thirty-two metres above ground level. You could also reach the test platform by external staircases.

The final days of August were taken up with intensive preparations for the test. Everyone had read the American news stories about the fault in the assembly of the atomic bomb's central unit and were ready for unexpected eventualities. The assembly was carried out in a special building, in strict accordance with Khariton's instructions. The assembly was successfully completed. The car containing the hardware was rolled onto the lift of the test tower. After the car was anchored, the rails disengaged, and the lift readied for raising the car, an unscheduled event took place: according to the regulations, the hardware was supposed to go up without people. Nevertheless, after the order had been given for the lift to go up, seconds before the motor started, Zernov hopped into the cabin. He went up to the test platform together with the hardware.

During the night of 28 August 1949 most of the participants in the test hardly slept at all. The explosion was scheduled for six o'clock on the twenty-ninth. The final night gave rise to anxiety for meteorological reasons as well. It drizzled, but cleared up a little towards morning. The sky remained overcast, but there were no visibility problems for the optical instruments poised to register the explosion at a radius of ten kilometres and more.

The control panel was located in a building ten kilometres from the tower. On the way from the epicentre to this building, the last safety barrier was three kilometres from the epicentre. Throwing a knife switch at that barrier automatically connected the generator circuit to the control panel. The final check. Everyone took their places. Shchelkin activated the control panel. All the operations which followed took place automatically.

Mal'skii, in a somewhat monotonous voice, counted off the final seconds: 'Eight seconds left, seven, six...!'

An explosion. A bright flash of light. A column of flame, dragging clouds of dust and sand with it, formed the 'foot' of an atomic mushroom. Kurchatov said only two words: 'It worked'.

Half a minute later the shock wave rocked the casemate. Even before it came, it was clear to everyone that 'it had worked'. What remarkable words these are: 'It had worked! It had worked!' Physicists and engineers, mechanics and workmen, thousands of Soviet people who had worked on the atomic problem, had not let the country down. The Soviet Union had become the second atomic power. The nuclear balance had been restored.

The power of the explosion turned out to be greater than had been predicted. This could be explained. Our discussions on pressure at the detonation wave front had not been in vain. The theoreticians had used the lower limits of experimental measurements in their calculations. The lack of precise data on the equation of state for plutonium also played a role. It had been assumed to be close to that for uranium.

On 25 September 1949 TASS carried the announcement that the Soviet Union had the secret of the atomic bomb:

> ...on 23 September President Truman announced that, according to US government information, an atomic explosion had taken place in the Soviet Union in recent weeks. Similar statements have been made by the British and Canadian governments...TASS considers it necessary to recall that as early as 6 November 1947, in a report given on the occasion of the thirtieth anniversary of the October Revolution, it was announced that for some time that had been no such thing as the secret of the atomic bomb. This announcement signified that the Soviet Union had already discovered the secret...
>
> As for the alarm over this which has been spread in certain foreign circles, there are no grounds for it. It must be said that the Soviet government, despite its possession of atomic weapons, still maintains, and in the future will continue to maintain, its former position on the unconditional banning of the use of nuclear weapons...

On 29 October the announcement was made of awards to the participants in the developing and testing of the country's first atomic bomb. This first award ceremony was especially memorable. It took place behind closed doors, without newspaper coverage. Soon after the November holidays Zernov phoned and asked me to come and see him at the Red House, and to bring Tarasov with me. When we entered his office, he rose ceremoniously from behind his desk, shook our hands and read out an excerpt from the award announcement. We had been awarded the title of Stalin Prize laureates and decorated with Orders of Lenin. Everyone was particularly impressed with the 'flying carpets'. That was the name people soon gave to the leather booklets with the emblem of the Soviet Union on the front. Their owners were entitled to unlimited travel free of charge on all forms of public transport throughout the USSR. The booklets were issued for life to the leading figures in the atomic project and to their wives. However, after Stalin's death they became invalid. The same awards were given to the members of Al′tshuler's division—Kormer, Krupnikov, Ledenev and others. In all, about a thousand people received awards. Directors were awarded the title of Hero of Socialist Labour. In Arzamas-16 the highest honours went to Khariton, Zel′dovich, Zernov, Shchelkin, N.L. Dukhov, and Flerov. In recognition of their special services Kurchatov and Khariton were given comfortable seven-seater ZIS-110 cars.[7] These gifts were to cause much trouble to their owners. The capricious ZIS-110s needed special heated garages, aviation fuel, special lubricating oils, and expensive servicing. Overworked scientists had no time for all this. In the late 1950s both cars were sold through second-hand goods shops to representatives of the Orthodox Church.

In a sense we became the first people in the country to supplant figures from the humanities. Boris Slutskii responded to the situation with the following lines:

> Somehow physicists are all the rage,
> Somehow lyric poets are not.[8]

And then in American newspapers reports began to appear of the achievements of American atomic scientists. Edward Teller had formulated the problem of creating a superbomb based on the fusion of hydrogen isotopes. Naturally, even after the first test, we retained an acute sense of responsibility. Our work continued.

---

[7]The others were given the smaller *Pobeda* cars. All received *dachas*.
[8]From Slutskii's poem 'Physicists and Lyricists' ('Fiziki i liriki', 1965).

The authors in 1933

The authors in 1978

Tsukerman and N. K. Reshetskaia in the X-ray laboratory, 1941

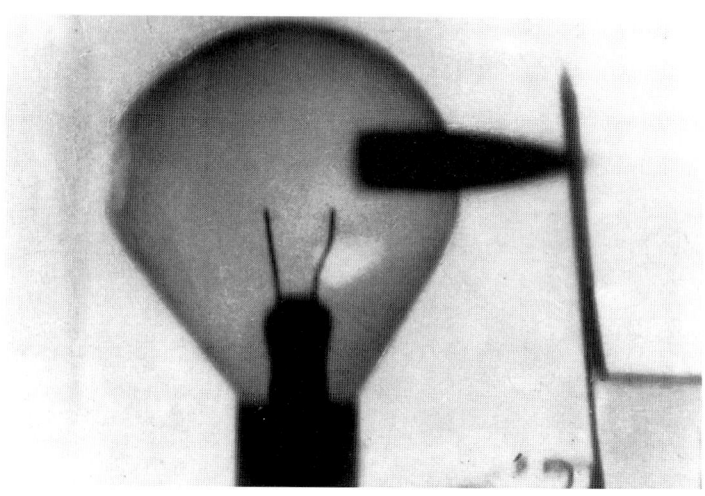

X-ray photograph of a bullet passing through a light bulb taken just before the glass has shattered, 1945

The Institute's first research laboratory was accommodated in this building in 1947

High-tension 500 kv. impulse radiography apparatus, 1946

The first purpose-built laboratory, constructed in 1948

M. A. Manakova assembles the charge for X-ray experiments

View of the town of Sarov in 1990

V.A. Aleksandrovich

L.V. Al'tshuler

A.K. Bessarabenko

A.A. Brish

V.A. Davidenko

M.V. Dmitrev

S.B. Kormer

V.V. Sof'ina

D.M. Tarasov

P.M. Tochilovskii

M.A. Kanunov

I.V. Kurchatov

Iu.B. Khariton

I.B. Zel'dovich

I.E. Tamm

A.D. Sakharov

P.M. Zernov

B.G. Muzrukov

Part 3

THE ONES WHO BEGAN

In this chapter we will talk briefly about the scientists, engineers, inventors and technicians who participated in the creation of the very first Soviet atomic weapons. Their contributions were in many cases decisive. Many of them are no longer alive. This places additional obligations on the surviving witnesses of their heroic deeds.

> They should make nails out of people like that.
> They'd be the strongest nails in the world.[1]
> *Nikolai Tikhonov*

VITALII ALEKSANDROVICH ALEKSANDROVICH

HE was born on 27 February 1904 in Odessa.[2] It is curious that on that very same day, on the banks of the Neva, in another port city, another man was born who was to play a large part in Vitalii Aleksandrovich's destiny—Iulii Borisovich Khariton.

They first met in Leningrad in 1931 when Aleksandrovich, a graduate of the Dnepropetrovsk Chemical and Technical Institute came to work at the recently formed Institute of Chemical Physics.

Here is how Khariton recalled their first meeting:

> In one of the rooms of the institute I saw a very large man with powerful hands soldering something over a glassblowing

---

[1] The final couplet of the poem 'Ballad of the Nails' ('Ballada o gvozdiakh', 1919-22) by Nikolai Tikhonov.
[2] This section is based on material supplied by E. and G.V. Aleksandrovich.

torch. I went closer. He turned out to be sealing off a test-tube filled with a yellowish liquid.

'What's that you have there?' I asked. 'Nitro-glycerine,' he answered. 'And you're not scared?' 'While I'm sealing off the test-tube, I hold it with my fingers below the level of the nitro-glycerine inside. Your fingers can withstand temperatures of over 70 degrees. That means that as long as I can stand holding the test-tube, the temperature of the liquid is much less than that. The temperature at which nitro-glycerine explodes is over 200 degrees.'

This was typical Aleksandrovich. His motto was always: 'If you have knowledge, there's nothing to fear.'

He loved risk, but he always combined it with caution based on knowledge. For him sport was an activity with a constant risk-factor. That was why, in his youth, he had been keen on all kinds of sport: swimming, rowing, sailing, skiing, boxing, shooting, mountaineering. It was clear that even in science he found a certain sporting element, a combination of risk, caution, knowledge and confidence in one's own abilities. All the same, in the last years of his life he overstepped the bounds of acceptable risk more than once. It is difficult to say whether this was because he underestimated the danger or because he overestimated his own strength, or whether it was a conscious choice on his part. The result was lung cancer, the tragic outcome of which was only postponed by radical surgery.

Several years after the death of Aleksandrovich, Khariton gave a speech at his own sixtieth birthday celebration. In it he recalled the people it had been his lot to work with and described Aleksandrovich as a past master, possessed of a range of striking talents, to whose hands and clear head the whole country was indebted.

Hands and clear head. A very accurate description. He had many childlike traits which determined his life-style: directness, irrepressible curiosity, a distinctive and unclichéd view of ordinary things, an originality of thought and an urge to try anything himself. All this allowed him to find unexpected solutions to the most varied of problems.

His skill with his hands was largely inherited from his father, a remarkably skilled physicist and technician who had done delicate virtuoso handiwork since childhood.

Aleksandrovich inherited his appearance from Zaporozhian Cossacks on his mother's side. It has to be said that his appearance in no way corresponded to the image of the intellectual which we get from literature. Tall, powerfully built, he had a muscular frame acquired through the hard physical labour he had had to do since the age of twelve. Aleksandrovich liked to demonstrate his physical abilities in a joking sort of way. He would lift the back of the Institute's *Gazik* and prevent it from moving.

His face was far from refined, with its great potato nose. Such faces can be seen in the crowd of Zaporozhian Cossacks in the famous painting by Repin.[3] At first glance you would never have suspected that he was a profound and original scientist.

The thick rimless glasses he wore did give his face a certain intellectual appearance. But these same glasses hid his eyes, which revealed their owner's true character. They were small and dark-brown and lent animation to his face, making it appealing and attractive. Like many strong people he was kind, but not soft; in questions of principle he made no compromises with his conscience. He took a paternally benevolent attitude towards junior members of staff. For these qualities, while still far from being an old man, he acquired the nickname *Batia* (Father), which stuck.

Everyone who knew 'Father' unfailingly remembers him on his heavy motorcycle with its sidecar. He would ride around on it winter and summer, in all kinds of weather, even after he acquired a *Pobeda* car. His massive figure seemed to merge with the motorcycle and reminded one of a latter-day centaur. It often happened, if I had to work late, that Aleksandrovich would sit me in the sidecar and drive me home at breakneck speed.

I remember some pranks, some harmless practical jokes, which broke the tension during a period of intense work, and various 'scientific' eccentricities. For example, he named his children using Mendeleev's Periodic Table as a kind of church calendar. Thus there appeared first Helium, then Rhenium and then Selenia. True, his son Konstantin, whose name was registered by his grandfathers without his father's knowledge, fell outside the bounds of the Table. He was to have been Tritium—Trishka for short. Aleksandrovich's eldest son continued this tradition by christening his own firstborn Ruthenium.

---

[3] The painting referred to is 'The Zaporozhians Compose a Letter to the Turkish Sultan' (1878-91).

Aleksandrovich's scientific interests were very varied, although he was particularly drawn to explosives. During his stint at the Institute of Chemical Physics he carried out his first scientific work on the self-catalyzing decomposition of nitro-glycerine. At the invitation of Academician L.V. Pisarzhevskii he returned to Dnepropetrovsk and worked there as a senior researcher in the Institute of Physical Chemistry until the outbreak of the Second World War. During that time he conducted a number of research projects on powder combustion speed, thermal decomposition of lead azide, and the explosive mechanism of hydrogen-oxygen mixtures. He also did work on corrosion-proofing aircraft engine pistons which had important domestic and military ramifications.

However, the most significant work Aleksandrovich did before the war, work which determined his subsequent career, was his obtaining heavy water through the electrolytic decomposition of ordinary Dnieper river water. The essence of this process lay in the fact that, during the electrolysis of water, part of it is enriched with molecules containing a heavy isotope of hydrogen—deuterium. The apparatus designed by Aleksandrovich consisted of a cascade of electrolytic baths, with an enrichment at outlet of 90-95%. His apparatus produced four cubic centimetres of concentrate per month, a figure which nowadays looks very modest indeed; this was, however the Soviet Union's first heavy water. The war interrupted this line of work.

During the war, because of his eyesight, Aleksandrovich served in a non-combatant unit with the rank of junior lieutenant-technician. When Laboratory No. 2 of the USSR Academy of Sciences was set up in Moscow under the directorship of I.V. Kurchatov, Aleksandrovich's pre-war research aroused new interest. In 1944, by order of the State Defence Committee, Aleksandrovich was recalled from service and put at Kurchatov's disposal. The new branch of physical chemistry which he instigated continues to develop to this day. Aleksandrovich learned that he had been awarded a Lenin Prize for this work while in hospital, just over two months before his death.

He died on 12 July 1959.

Lev Vladimirovich Al′tshuler

Lev Vladimirovich and I have been close friends for over sixty years now. As I have already mentioned, our friendship began long ago, in 1928, when we were at school together; since then, except for brief periods when circumstances separated us, we have travelled through life together. We have worked together and our interests have been closely intertwined. Ideas that have occurred to one of us have always stimulated a lively response in the other.

His mind-set was much more that of a researcher than mine. When he graduated in 1936 from the physics department of Moscow University, he had a complete grasp of mathematics and a profound understanding of experimental physics. As he himself used to say, he considered it his mission in life to explain experimentation to theorists and theory to experimenters. I once happened to be present at a conversation between Zel′dovich and the mathematician S.K. Godunov, who opened by saying:

'I've come to share with you the rubbish I've just had from the young physicist Lev Al′tshuler. Two days ago you and I came to the unanimous conclusion that the problem I was telling you about had no unequivocal and straightforward solution. And he's solved it beautifully.'

From Al′tshuler's first days at Arzamas-16 his sharp, analytical, inventive mind marked him out for one of the very front ranks of the experimental physicists there, ones whose competence included gas dynamics, mathematical analysis, and other related disciplines.

Al′tshuler's confidence in the correctness of his own judgements and a sense of a certain impunity in his dealings with others often created difficulties not only for himself, but also for his friends. For example, in spring 1951, during the re-attestation of our leading research scientists, the chairman of the attestation board asked Al′tshuler how things stood with his political education.[4] Like the majority of physicists Al′tshuler took a negative view of Lysenko's 'ideas' concerning the possibility of regenerating oats into wild oats and of his criticism of Mendel's laws. He particularly objected to Lysenko's anti-materialist approach to questions of classical genetics and heredity. At that time, circumspect physicists gave evasive answers to such

---

[4]Attestation and periodic re-attestation of professional personnel was an external review process which was required by the Soviet régime and designed to monitor professional qualifications and performance, as well as ideological 'fitness'. Careers could be obstructed or wrecked at this periodic formal hurdle.

questions, but Al'tshuler embarked on an attempt to convince the commission of the materialist essence of genetics.

A few days later came a strict directive from one of Beria's assistants, P.V. Meshik to dismiss the 'Weissmanist-Morganist' Al'tshuler.[5] During these critical days for Al'tshuler, Avraamii Pavlovich Zaveniagin came to Arzamas-16. At midnight I managed to arrange a meeting with him. I told him in detail about Al'tshuler's proposals and research, pointing out the irreparable damage his dismissal would cause to our work. At the end of our talk Zaveniagin asked: 'Do the other scientists here share your view?' 'I haven't talked to them, but I presume so.'

Next morning Zababakhin and Sakharov paid a visit to Zaveniagin. It looked as if everything had ended favourably. But fairly soon afterwards[6] Khariton had to appeal directly to Beria on the same account. He told us:

'I rang Beria, told him Al'tshuler was very much needed for our work, and requested that he be left in post. Beria again asked whether he really was very much needed. I confirmed that he was. Then Beria agreed to leave Al'tshuler in post.'

In a mock poem read at one of his birthday celebrations Lev Vladimirovich was dubbed 'Levka the Dynamite Man'. The name stuck.

Many physicists considered us general trouble-makers. On all issues that affected our interests we took up what were, for the time, extremely active and principled positions, bordering sometimes on the aggressive. In 1953 Zel'dovich wrote a very revealing inscription in an offprint of his monograph which he was giving us: 'To the Robber-Brothers, from the author who has not yet fallen victim to them.'[7]

Al'tshuler's subsequent career resembled an obstacle course. In 1956 he was expelled from the *nomenklatura* and removed from his post as scientific director of a large department at Arzamas-16. This happened after his speech at a Komsomol

---

[5]A slur applied to geneticists by Lysenko and his state-sponsored supporters. The name derives from two founders of modern genetics, the German August Weissman (1834-1914), and the American Thomas Hunt Morgan (1866-1945).
[6]According to Al'tshuler himself, the appeal to Beria took place one year later.
[7]A reference to the short narrative poem 'The Robber Brothers' ('Brat'ia-razboiniki', 1821-22) by Aleksandr Pushkin.

debate on Dudintsev's novel *Not by Bread Alone*.[8] Having stumbled across the debate by chance, Al'tshuler gave an impromptu exposition of his stance on social issues. He spoke of what he called the 'unidirectional conductivity' of Soviet society (i.e. from the top down) and, furthermore, recommended adopting the Yugoslav model. In subsequent years Al'tshuler's independent position on many issues prevented him from taking part in elections to the Academy of Sciences. Finally, in 1969, he left Arzamas-16.

A kind heart and a fundamental urge to help others—that is the basis of this complex character. Many people recall Lev Vladimirovich with gratitude for the active help he gave them in difficult situations.

The Soviet school of explosives occupies a leading place in world science. Al'tshuler's contribution to this branch of science and technology is difficult to overestimate. He demonstrated convincingly how to obtain and study megabar pressures by comparatively simple means. In recognition of this work Al'tshuler received two State prizes, one Lenin prize, and three orders of Lenin. In 1991 American colleagues awarded him an American Physics Society prize. In Edward Teller's estimation, Zel'dovich and Al'tshuler are the two scientists who have contributed more than anybody to the discovery of a new field of research—the physics of high energy densities.

ALEKSEI KONSTANTINOVICH BESSARABENKO

OUR acquaintance began with music. The night before we had all agreed to look in on Bessarabenko at ten the next morning. He lived with his wife and three sons in a hostel. We entered the building's long corridor. We asked a young man who was rushing past us whether he knew where the Bessarabenko family lived. 'Can you hear that singing? That's where you want.' At the end

---

[8] A talisman of the post-Stalin 'Thaw'. Published in 1956, it deals with an inventor's struggle with the bureaucratic system.

of the corridor a pleasant male voice was singing a popular aria from the operetta *Rose-Marie*.[9]

A few moments later we met the owner of that lyrical tenor voice, Aleksei Konstantinovich Bessarabenko, along with his wife Aleksandra Aleksandrovna, a pleasant, grey-eyed blonde.

The son of a naval technician, Aleksei had begun working at the age of thirteen in the shipyard at Sebastopol. At first, the work was straightforward: together with other boys he descaled ships' boilers. The job of boilermaker called for small stature, since you had to crawl into the boiler through relatively narrow openings. Their father would tell both little Aleksei and his brother Nikolai: 'A man without skills is not a man but an oversight.' Aleksei remembered this lesson for the rest of his life. Working in the shipyard brought out an important trait in his character—limitless respect for work and working people.

After graduating in 1935 from the Kirov Urals Industrial Institute, Bessarabenko worked at an artillery plant in Perm, first as a production engineer, then as a foreman, then as a divisional head, and finally as chief of production. His talent and tremendous appetite for work as a manager ensured the success of his team. 'The section headed by Bessarabenko from 1939 to 1941 consistently overfulfilled its production targets,' we read in one of the documents preserved from that time.

In the summer of 1947 Bessarabenko was appointed director of mechanical production at Arzamas-16. The work was difficult. He had to create an efficient and harmonious team in a new environment, a team which would face complex tasks and be given tight schedules. In this situation Bessarabenko demonstrated his organisational talent. 'Overcoming difficulties, breaking through barriers—that's when he's in his element,' said one of the senior employees at Arzamas-16. Workmen, service personnel, shop foremen, and scientists all maintain that Bessarabenko always found the right approach to take with them. However busy or tired he was, however late it was, Bessarabenko never refused to see anyone and always tried to understand people and to help them. For failure to fulfil work obligations he could issue

---

[9] Operetta by Rudolf Friml (1879-1972), libretto by Oscar Hammerstein and Otto Harbach. Produced in New York in 1926, it was staged in both Moscow and Leningrad in the same year. In the 1930s it was permanently in the repertoire of the Moscow Theatre of Operetta.

a fierce reprimand. In character he was explosive, but fair. The ability to see the human being behind the assignment was one of the basic traits of that character.

He worked very long hours, often days and nights. His staff, knowing that there were not enough manual or skilled workers, would often remain at work with him long after the end of their shift. 'One night,' recalled Ananii Il'ich Novitskii, 'Bessarabenko came into the division I headed. Suddenly I felt ill with exhaustion. A doctor was called. He looked me over and said I needed to rest. Bessarabenko offered to take me home. I looked at him. His face was tired, but his gaze was alert, and I declined and said there was work to do. To be honest, my conscience wouldn't let me. After all, I knew that Aleksei Konstantinovich himself was spending literally days and nights on end at the plant.'

In 1952 Bessarabenko became deputy director of Arzamas-16 and, in 1956, chief engineer.

A heart attack cut short this brilliant life on 24 October 1960.

> And no-one will ever forget him
> In his unfastened coat,
> Headlong and determined as the wind...
> Yes, it's a pity we shall not see him again,
> But even so we're grateful
> That we were alive with him.

So wrote Galina Bednova, the senior construction engineer at Arzamas-16 of Aleksei Konstantinovich Bessarabenko.

ARKADII ADAMOVICH BRISH

HE was born the son of a teacher in Minsk, the capital of Belorussia, on 14 May 1917. Born in the same year as the October Revolution, for the first twenty-four years of his life he thought that this date ought to bring him luck. But Fate was harsh to him. In June 1941 the Germans took Minsk, and Brish, a member of the Komsomol who had recently graduated from the physics department of the Belorussian State University, became a partisan. He took part in all the battles in which his brigade fought, and also went on missions with partisan groups in which he showed himself to be both bold and decisive. For his role in the partisan movement Brish was given the medal 'Partisan of the Great

Patriotic War, First Degree' as well as the Order of the Red Star. After he was demobilized in October 1944, the headquarters of the partisan movement assigned Brish to the Academy of Sciences to continue his scientific work.

We met in the first year after the war, soon after he joined the Institute of Machine Science of the USSR Academy of Sciences, where I was working. In those far-off days there was still no personnel department in our organisation, and staff were selected by the laboratory chiefs. At the end of our first conversation I asked Brish: 'Have you any regrets about landing this assignment?' 'Taking part in work like this,' he answered, 'is the same as fighting for republican Spain against Franco and his fascists. I would lose all respect for myself if I declined your offer.'

The skills which Brish acquired as a partisan came in very handy. There were many occasions when they got us out of difficult situations.

Brish was a very colourful figure. In 1946, when Arzamas-16 was still under construction, he was twenty-nine. He invested all his inexhaustible, irrepressible energy in production work. He was a man you could trust with the most complex and responsible tasks. It was a time when not only new methods and new instruments, but also new nicknames, were being invented. Someone proposed a new name for one unit of production activity—'one brish'. This was an unattainable value. Most applications called only for 'millibrishes' and 'microbrishes'. In Part 2 of the present work, we discussed the 'Brish effect'—the phenomenon of high electrical conductivity in explosion products and dielectrics under compression measured in megabars. The phenomenon was discovered by Brish and his colleagues in 1947.

A scene comes to mind from out of the depths of my memory. It was late evening. In the laboratory we were awaiting the return of Brish and M.S. Tarasov. The first experiments with electromagnetic measurements of mass velocities of explosion products from secondary explosive substances were under way. They appeared about eleven o'clock in the evening in snow-sprinkled sheepskin coats. Both of them looked handsome and a bit excited. Mikhail Semenovich Tarasov was a short, blue-eyed, fair-haired man with a military bearing—during the war

The ones who began

he had been a radio operator aboard the *Baku*.[10] Under his unfastened sheepskin coat you could see a naval jacket. Brish was also fair-haired, with grey eyes nearly the colour of steel. The oscillograms immediately became the subject of heated discussion. I kept repeating to myself that I must fix all this in my memory: the late evening, the condenser flash unit in the corner, these people, my closest friends, with ardent hearts and huge reserves of energy.

At my lectures and talks on scientific ethics, people would ask me what qualities a scientist needed to possess in order to achieve success in a short time. Of course, he must have professional knowledge, and must know how to use contemporary equipment and computer technology. But all of this alone is insufficient. The ability to impart to one's fellow workers enthusiasm and the will to succeed is very important. Without that will, without that belief in success, no victory is possible. Brish possesses these spiritual qualities in full measure.

When Brish turned his attention to extracurricular activities, his department really came to life. He had everyone running the hundred metres in the stadium, including the departmental chief. Tall and gaunt, Brish is still a keen downhill skier. Everyone found working with Brish easy and interesting. Clapping a lab assistant warmly on the shoulder and smiling, he would usually add: 'Only you can do it as quickly and as well as that,' even though he might have been talking about the simplest of tasks.

From 1948 onwards Brish became the most active participant in the development of an impulse neutron source, a course of action proposed by myself and Zel'dovich. A group was organized within the department he headed to carry forward this work.

In 1952 work on an external neutron source was given the go-ahead by the Scientific and Technical Council, chaired by Kurchatov. The Council took the decision to test this source in the automatic detonation device of the 1954 atomic bomb. In order to carry out this instruction, Khariton, who was a keen supporter of the idea of impulse neutron detonation, requested the Prime Minister, Georgii Malenkov, to put at our disposal the Moscow automatic aviation systems factory. As early as 1953 this

---

[10] It was aboard this ship that Kurchatov and N.S. Khrushchev, then First Secretary of the Communist Party of the Soviet Union, travelled to England in 1956.

factory had produced the first automatic detonation device which used an external neutron source. From 1955 this factory was put under the same ministry as us. That same year Brish and a group of colleagues moved to Moscow, where he became the permanent scientific director of this factory. In 1983 he was awarded the title of Hero of Socialist Labour.

To this day Arkadii Adamovich combines a sense of enormous responsibility with optimism and humour which his colleagues find infectious. When I think of him, I still see the Brish I knew during the first few years of our work: fair-haired, well-built, ambitious, and bold. It seems as if the years have roared past, preserving him as he was.

Viktor Aleksandrovich Davidenko

He and I met during the war, in 1943, in Kazan. He was then a research worker at the Leningrad Physics and Technology Institute, which had been evacuated from Leningrad, while I was head of the radiographic laboratory at the Academy of Sciences Institute of Machine Science. At this time he was working under Kurchatov at the recently formed No. 2 Laboratory. The whole of his subsequent life was bound up with nuclear science and technology. After Kazan there was a five-year interruption in our meetings, but contact was renewed, on another meridian, in January 1948.

Davidenko was a cheerful, active man, a lover of life, nature and animals. He had two hunting dogs. Whenever he was away for long on business, he left the dogs with us. Many people remember his pointed epigrams at the expense of Zel'dovich, Brish, and Sakharov, and the apt nicknames he gave them.

As I write, I still cannot believe that all this was in the past. Time and death are merciless. For Davidenko the hour came on 15 February 1983. And yet...'Blessed is he who visited this world in its fateful minutes...'[11]

In our archives we have kept N.D. Iur'eva's memoirs of her work under Davidenko, as well as Davidenko's memoirs of his contacts with Kurchatov. These stories follow below, with slight abridgements. Nina Iur'eva writes:

---

[11]Lines from the lyric 'Cicero' ('Tsitseron', 1830-31) by Fedor Tiutchev.

The ones who began

The years I worked under Davidenko, 1948-50, remain in my memory, and the memory of many of our colleagues, as the most interesting, the most fruitful, and the most happy of years. The reason for this lies not only in the interesting, critical, and responsible work we were doing, but also to a great degree in the personal charm of our director. He seemed to be charged up with an inexhaustible supply of energy, cheerfulness, humour, zeal, and optimism. We were always working to deadlines and you could find Viktor Aleksandrovich everywhere at once in our department. He would be feeding ideas to some, encouraging others, dreaming up ingenious constructions with a third group, and helping a fourth group to discover the reason for some inexplicable deviations in their results.

At that time we were short of instruments, materials, and equipment. Many things came from Moscow in very limited quantities. Supply problems weighed heavy on the shoulders of the research directors, but Davidenko would always find a clever way out of the predicament.

Many years later, reminiscing pleasurably about our work in those extraordinary years, Davidenko told us, with his usual humour, how he got out of perpetual time trouble: 'For example, I would say: Nina, this component will have to be acid-dip pickled for half an hour, and then rinsed for an hour in distilled water, but I want it ready in five minutes!'

Davidenko loved music, especially classical music, of which he had an excellent knowledge. While he was working he would often hum some simple melody and make up comic verses to it. To the tune of 'La donna è mobile' he used to sing:

> Little teapot with little lid,
> Little lid has little bump,
> Little bump has little hole,
> Little hole lets out steam.

How happy he was when we finally had our first hardware assembled! And, despite all our difficulties, assembled on time! Assembled despite the fact that our metallic vacuum unit had sprung a leak, which we had had no time to fix. The quartz ampoules, made for us in Moscow, had cracked after the first experiments and we had carefully smeared black picein over the cracks. We compensated for the short time available by working at night. And then everyone in the laboratory ran into the darkest room to admire the fruits of our protracted labours. They glowed in the darkness with an extraordinary, unearthly rosy-blue light, and we watched them with bated breath...This was in August 1949.

Davidenko and I met for the last time in 1977. He came to attend the anniversary of one of the departments and on the Sunday he made visits—as though taking his leave of them—to all the places he remembered, such as the house where he used to live. He visited the family of Aleksandrovich, whom he remembered with great fondness. Davidenko spent the second half of the day with us and we were glad that our children appreciated in him above all the same characteristics that were so dear to us: his simple and natural way with people, his openness, his ability to hold a conversation on a theme which especially interested his interlocutor, his great erudition and his profound respect for people.

Davidenko entitled his memoirs 'How I Became a Haystacker':

Probably no-one could talk about Igor Vasil'evich Kurchatov and do justice to him. He was a man on a big scale, which makes it difficult to avoid exaggerations and additions. In any case, nothing human was alien to him. As you know, Kurchatov loved to give nicknames to people. It was done in a friendly spirit, and gave no offence. As a rule, the nicknames arose spontaneously, but, one way or another, they characterized the person or their actions. I don't know which of us held the record for nicknames devised by Kurchatov, but I was clearly close to it. This is how I now remember my last christening by Kurchatov. At the time we were on a long business trip and living in a hotel on the bank of a broad and deep Siberian river, the Irtysh. We were working very hard. Late one evening, after repeatedly taking the same measurement ('for the statistics'), I said to Kurchatov: 'Isn't that enough self-indulgence? Everything's repeating itself nicely. It's clear everything's all right; the statistics are fine and we should have been in bed a long time ago.'

'All right. Let's just give it one more 'dzyk' (since his Leningrad days, when taking measurements, Kurchatov would always precede starting the stopwatch or flicking on the starting tumbler with the command: 'Ready...dzyk', upon which he pressed the button).

The ones who began

I was already fed up with 'dzyking', and tried to talk Kurchatov out of it: 'We've already done twenty dzyks and the result's the same each time.' 'What's up? Don't you want to give it another try? You can catch up with your sleep tomorrow.'

'How can I, if I have to be back at work by eight, having eaten some sterlet and made my way here?'

At this point 'the Beard'[12] took his beard in his fist and, drawing out the vowels, intoned: 'But we've got to-mor-row off.'

'How do you mean, got tomorrow off?' I said in surprise, since the concept of a day off hardly accorded with either the situation or our working habits.

'Just that. We've got tomorrow off, for (the word *for*[13] was somehow underlined) certain sons of bitches have cocked things up and failed to send us our bits and pieces.' After that we did another thirty or forty 'dzyks' until, finally, Kurchatov delivered himself of his usual: 'Get some rest.'

I'd known the significance of these words for a long time, and when I heard them, I at once decided that I'd been well and truly had and made no further mention of the day off, so as not to seem ridiculous. But the Beard asked me: 'So what are you going to do tomorrow?'

This gave me hope and I asked: 'You mean we really have got tomorrow off?' 'Don't you understand Russian? I told you—day off to-mor-row.'

'Cross your heart? Swear to God?'

Having started to believe in the possibility of a day off, I asked: 'Can I go off somewhere?' 'Go where you like.'

'Where you like' had been a settled issue for some time. I wanted to go duck-hunting in the nearby marshes.

'In that case I'm going hunting.' 'Off you go, but don't forget to bring back a quarter of a duck for each barrel.'

This was another gibe at the fact that I had a four-barrelled shotgun. I got changed quickly and hurried off to find the driver the *Gazik* assigned to us. He wasn't a hunter himself, but received my proposal with alacrity (anything was more interesting than sitting in barracks). We made a quick stop to pick up the four-barrelled shotgun and

---

[12] i.e. Kurchatov. See 4.1.

[13] Kurchatov uses the slightly archaic word Russian word *ibo*, presumably for emphasis.

the ammunition, and an hour later we were already hurtling down the road towards the marsh in order to get there by dawn.

There were ducks everywhere you looked, but they kept rising from the water a long way off and I kept missing spectacularly. It was a good job that no-one but the military driver saw me. By about eleven o'clock I'd got two ducks, that is to say half a duck per barrel, all my cartridges were spent, and I was on my last legs. The driver and I climbed onto a heap of dry reeds and slept sweetly for about two or three hundred minutes.

In the evening, when we drove up to the hotel, a major came up to us and said that the Beard had been looking for me all day and had been giving them hell all day because they couldn't find me. Dressed just as I was, I went up to Kurchatov's room and found him in excellent spirits. When he saw me, he at first tried to assume a glowering look, but such was my appearance that he could not manage it, and his eyes could not repress their gleeful slyness.

'Where the hell have you been? The whole garrison's been looking for you all day. Where were you at lunchtime?'

'In a haystack, of course, where a hunter should be at lunchtime.'

'You wretched haystacker! People are working and you're cooling off in a haystack.'

Still not realising that another christening had taken place, I told Kurchatov that the day before he'd given me leave to go hunting, and then began to explain to him in detail why a hunter had to be in the marsh at dawn but asleep in a haystack during the day. Kurchatov was completely happy with this explanation, but he continued ribbing me: 'PMZ[14] is a fisherman and went fishing, but we found him straightaway.'

'You think that was a big achievement? He was probably sitting opposite the hotel, half-asleep over his fishing-rod. If he'd gone off further, you wouldn't have found him either.'

'Well, from now on, haystacker, you'll be doing your hunting near the hotel!'

'But you said we'd got to-mor-row off,' I said, attempting to defend myself.

---

[14] i.e. Pavel Mikhailovich Zernov.

The ones who began

'So you think it's a bad idea to confer on a day off? Sometimes you have to confer, even on days off. It's sometimes useful to confer even with a haystacker.'

'All right. Let's confer.'

'It's too late now. We've already conferred without you. Go and get washed, then come and have some sterlet. Hand your ducks in to the kitchen. We'll try them tomorrow evening.'

I went and had a shower, having firmly grasped the idea that on long business trips you don't get days off, even if they swear to God that you do, and that from that day onward my name was 'haystacker'.

It was with this title that I took my leave of Kurchatov for the last time; my greatest misfortune is that now there is no-one to give me a new name.

MIKHAIL VASIL'EVICH DMITRIEV[15]

BROAD-SHOULDERED, well-built, with an open, manly face, he was a favourite among the staff. He would look at his interlocutors with serious, kindly eyes. And at the same time there were in these eyes, somewhere in the very depths, almost invisibly sly glints. You could see them easily when he met people he liked. Then he would take pleasure in joking with them and his eyes would gleam slyly. He did not like unprincipled people and did not hide the fact.

He was a marvellous storyteller. Many extraordinary and comic things had happened either to him or to people close to him (maybe he made some of it up). He would tell stories with humour, and laugh along with his listeners, of whom he was never short. In 1948 he was thirty. By that time he was already a family man: his son Sasha was ten, his daughter Dina about seven. He had marched a good many miles during the war, by the end of which he had reached the rank of captain-engineer. He graduated from the Military Chemical Academy.

Dmitriev was an first-rate chemist, his speciality being radiochemistry. He had an excellent grasp of technology, which was quite rare for a chemist at that time. With his own hands he made almost invisible components for electrolyzers, in which he

---

[15] This section includes material supplied by N.D. Iur'eva.

carried out experiments with weightless quantities of certain elements. He assembled models of various mechanisms—turbines and steam boilers—out of small components he himself had made by hand. His wife, a teacher of physics, liked to demonstrate these mechanisms during physics lessons with her older students.

For Dmitriev's radiochemical work two protective concrete walls and an exhaust hood were installed in a little room in the laboratory building. The room also contained, just, a workbench for electrolyzers, a cabinet for reagents, and a tiny desk. It was there that he worked with radium, extracting from it the first quantities of polonium. On Khariton's instructions, supplies of radium were delivered to us from various cities in the Soviet Union. The experiments would last for hours, sometimes even for days. After he had finished all his preparations and turned on the electrolyzers, Dmitriev would do the rounds of the other rooms. He only had to appear somewhere for all his colleagues to start vying for his attention. One would want to consult him about work, another to boast about his results, while another would simply want to talk to a pleasant and intelligent man.

It very soon became clear that work in such conditions was unthinkable and so a small Finnish cottage not far from the laboratory wing was fitted with equipment for Dmitriev's work. His first assistant was Liubov Moiseevna Brish (the wife of Arkadii Brish), a tall, attractive woman, always well-groomed and tastefully dressed. With a higher education in the humanities, she worked as a laboratory assistant, never shirking even the most dangerous and dirty work. During the wartime occupation of Belorussia she had had to work as a farm labourer until she managed to escape to join the partisans,

Dmitriev was always very willing to help anyone at work who needed him, and he willingly shared his knowledge and experience. I myself turned to him many times for advice and assistance. When we needed targets containing hydrogen isotopes, in a matter of days he suggested and implemented a method for making them, based on metallic hydrides. A few days before his death he called me to his hospital bed and suggested using a chemical compound for the measured introduction into gas of carbon monoxide, ammonia, and other gases. For decades after his death we were still using his suggestions in our work.

Dmitriev was one of the most industrious, modest, and selfless researchers in our department, always willing to carry out any difficult or dangerous task himself. There was not a trace either of careerism or obsequiousness about him. He treated everyone with equal kindness and particularly appreciated and respected people who were honest and hard-working. A talented chemist, he was one of the first to take up the new field of the chemistry of radioactive materials.

Here is how his close friends, Galina and Aleksei Voinov recall him:

> We met him in the early 1950s. At that time Arzamas-16 was growing rapidly. A single batch of appointments, for example, brought in eighteen people at once, all of them well-trained and self-confident physicists from the first post-war intake of the physics department at Moscow University. Dmitriev and Aleksandrovich immediately christened us 'the rhombic squad'. With characteristic good humour they were alluding to the silver rhombic badges worn at that time by university graduates (other institutions of higher learning did not have badges and secretly envied us), and to a serious defect, from the point of view of experimenters of the 1940s and 1950s, in university education—the fact that none of us had ever seen a machine tool or been trained in the rudiments of designing, let alone building, laboratory equipment, which neither wartime nor post-war industry produced. For all our theoretical background in a broad range of issues in nuclear physics, we lacked both knowledge and experience of the construction and manufacture of complex physical installations and instruments, such as accelerators, detectors, radiation sources, reactor and separation units, and so on.
>
> The older generation, although the difference in years between us was not so great, would often rib us rhombic nuclear physicists and play little practical jokes on us, whilst taking upon themselves the task of completing our education.
>
> Our friendship with Mikhail Vasil'evich was, unfortunately, brief. He soon moved away to a new job.
>
> Dmitriev belonged to the category of highly talented skilled craftsmen. Specialists in the field know that the accuracy of laboratory experiments involving neutrons largely depends on the power and dimensions of the neutron source: the greater the power and the smaller the dimensions, the better. Dmitriev decided to satisfy a long-standing wish of neutron specialists by developing and building a miniature neutron source of great power.
>
> It is difficult to understand how he managed to solve this complex problem at the dawn of nuclear technology in

this country. This was in the early 1950s and it took more than thirty years for technological progress to surpass the achievement of Mikhail Vasil'evich.

The talent of the technologist and the skill of the experimenter were combined in him with a boyish sense of adventure, with great physical strength, and with tremendous spiritual warmth. Once he horrified the administration when he jokingly took on two of Kurchatov's strapping bodyguards at wrestling. Another time he won a bet before a large group of witnesses by wriggling out of a building in the middle of winter through a small window which looked scarcely big enough for a cat. A wonderful, kind and enchanting man!

His rare talent and abilities were cut short by his early death. He had one stroke of bad luck, but radioactivity allows no second chances. He died of liver cancer in 1962, at the age of forty-four.

Samuil Borisovich Kormer

I took a liking to the young lieutenant the first time I met him. Sometimes you meet a person, have a talk with them, but, in order to get an idea of their professional calibre, you have to eat a pound of salt with them, as the Russian proverb has it. Sometimes it is like love at first sight—you talk for half an hour and you're friends for life. Such was my first meeting with Samuil Borisovich Kormer. It took place in summer 1946 in one of the Moscow research institutes of the Ministry of Munitions. Kormer was not yet twenty-four. He had been seconded to Arzamas-16 by the Artillery Academy to do his pre-diploma practical work and to complete his diploma research on the mechanism of cumulative charges. Determining the characteristics of such charges requires photochronographic technology, and at that time there was practically none in the Soviet Union.

The word 'photo-chrono-graph' contains three independent linguistic roots, which can be translated as 'light-time-writing'. Our first photochronographs, built out of secondary materials, were far from perfect. Nevertheless, Kormer set to work with them with all the fervour of youth and was one of the first researchers to determine the speed stream of small cumulative charges. These experiments measured the time it took for the cumulative stream to pierce various barriers. In this first piece of independent research Kormer's most important qualities were already made manifest—a high degree of professional activity, the ability to set himself a clear-cut goal and to succeed in spite of a mass of difficulties.

Khariton takes up the story: 'In 1946-47 I led a seminar in the Institute of Chemical Physics of the Academy of Sciences, at which various questions concerning the combustion and detonation of explosive substances were examined. Once, after the seminar, a young serviceman suggested that he should give a paper on the functioning of cumulative munitions. I agreed. During the paper it became clear that this man knew his subject very well, and I included his name on my list of staff for our future institute.'

Head-hunting Kormer turned out to be no easy matter. By this time he had successfully graduated from the munitions faculty of the Dzerzhinskii Academy and had been assigned to a defence plant not far from Moscow. There, the knowledge and professional abilities of the senior lieutenant-technician were soon recognized and they were reluctant to let him go. Khariton had to put some pressure on the military, and in August 1947 he joined our staff. The next few months showed how wise this choice was.

Together with him, we continued perfecting our methodology and technology for photochronographic research into the visible characteristics of explosive processes. Many of the principles and technical solutions which Kormer proposed and implemented in 1947 are still being used without significant modifications to this day. It is impossible to overestimate the role of this technology, in the development of which Kormer played such a decisive part. The same period saw pioneering work in the interaction of strong waves in various media. From the mid-1950s onwards Kormer became one of the leaders of a new branch of research which led to savings of hundreds of millions of roubles. For his accomplishment of this crucial government task Kormer was awarded a Lenin Prize in 1959.

The inimitable charm of our work during the second half of the 1940s is receding ever further into the past. But when you recall Samuil Borisovich Kormer, a flash of light seems to illuminate in the memory numerous remarkable episodes which took place during our work together. It was November 1947. Kormer and I were working in one of the protective structures. Our instruments had just registered optical phenomena during the explosion of a large-scale charge. Kormer was developing the film.

He appeared, a cheerful figure, in the doorway of our improvised darkroom. 'Look, here's our first rastergram!' he told everyone in a ringing voice. Two weeks later Kormer had his twenty-fifth birthday. During the celebration, which was almost a family affair, people continued to talk shop.

The basic scientific tendency, to which Kormer dedicated his whole life, lay in determining and studying the properties of substances, both solid and gaseous, under extreme conditions—high densities, pressures and temperatures. Kormer always preferred optical research methods, which gave the clearest impression of the phenomenon under study without, in most cases, influencing the course of the basic process involved. Under his guidance the temperature of solids under shock compression was measured, and unique data obtained on the structure of shock wave fronts through the reflection of light.

Kormer's preference for optical methods probably accounts for his interest in the possibility of using lasers for the compression of metals, and for the enthusiasm with which he set about this work, infecting both his colleagues and his superiors with the same enthusiasm. The work was done with typically Kormerian tenacity. Once he had set himself a goal, he pursued it with enormous inner strength, like a heavy tank, taking everyone with him, invariably overcoming every obstacle in his path.

Now let us look in on the conference hall of the division which Kormer headed from the outset. The room is full to capacity. Today they are discussing the first experiments involving laser-assisted thermonuclear synthesis. Once again there is the problem of light and time. But how far they have come from the problems which we faced thirty-five years ago. Then we were dealing with microsecond intervals. Now there are laser flashes lasting from 0·3 (a third of a millionth of a second) to 0·5 nanoseconds. This is not a beam but a 'fragment' of light only ten to fifteen centimetres in length, something like a light dart. A number of such darts fired from different sources simultaneously penetrate a ball of hydrogen isotopes some tenths of a millimetre in diameter. This is an incredibly difficult task. A large group of scientists and engineers, carefully put together by Kormer over many years, is working on this problem.

Concern for his staff—their daily living needs, their scientific work, their participation in institute meetings, and in seminars and conferences at national level—this was another of Kormer's positive characteristics. The result of this sort of relationship between director and staff was a stable core of highly qualified and reliable researchers who have been working harmoniously together for many decades.

'It was 1981. There was a big international conference going on,' recalls V.K. Chernyshov, a doctor of physics and mathematics. 'A large lecture theatre. The screen goes dim. A paper is coming to an end. The chairman announces the title of the next one to be given in this room and the title of one to be given by Kormer in the room next door. But what's this? Why is there such a commotion? Everyone is getting up and heading for the exit. They're going to the room next door, where there are Austrians, Japanese, French, Americans, and "sundry Swedes".[16] The paper begins. In the front row two elderly people are listening very attentively. One of them is a famous professor from Leningrad, from the school of Academician Ioffe. The paper is drawing to an end and I see the beaming face of the old professor. He turns to his neighbour and says with feeling: "Yes! That was in the best traditions of the Soviet school of physics!"'

Study of the equation for the state of hydrogen under superhigh compressions, the creation of a superpowerful nuclear unit, capable of producing the most powerful laser beam to date, experiments in laser-assisted thermonuclear synthesis—these are the pieces of original research conducted under Kormer's supervision or with his direct participation. In recognition of these achievements Kormer was elected a corresponding member of the Academy of Sciences in 1981.

Time imperceptibly imposes its limit on every human lifespan. Kormer succumbed to a heart attack at the height of his creative powers. He died on 10 August 1982, three and a half months short of his sixtieth birthday. But during his three and a half decades at Arzamas-16 he did so much towards solving our most crucial physical and technical problems that only a very few researchers could rival him.

---

[16] A quotation from Maiakovskii's 'Verses about my Soviet Passport' ('Stikhi o sovetskom pasporte', 1929). The phrase is used jocularly to refer to any representatives of small Western European nations.

Vera Viktorovna Sof'ina

A SHORT, plump, very lively woman, with a severe hairdo emphasizing her high, open brow. Vivacious, greenish-brown eyes. Often her whole face would be illuminated with a slightly ironical smile. That's how she arrived at Arzamas-16 in 1947 and that's how I remember her. At the time we were all young and Vera Viktorovna's age put us off at first: she was fifty. However, through her energy and youthful enthusiasm she soon proved that age is no handicap in our work. Sof'ina turned out to be a serious researcher and a person of extraordinary abilities and knowledge. Science, especially experimental physics, was her lifelong passion. But the path she took to reach it was not easy. Her native city was Saratov. She was born there in 1897, and she completed her grammar school education in 1913 with top marks in all subjects. She studied painting at the Bogoliubovskii School of Art. It was there that she met her future husband, the artist Aleksandr Petrovich Sof'in. In 1919 he was called up into the Red Army. Sof'ina went with him. In the army she taught literacy skills as part of the *likbez* programme. She was thirty-five when she graduated from the physics department of Saratov University. Before she joined us she worked for fifteen years as director of the radiographic laboratory of a factory which made batteries.

The death of Aleksandr Petrovich at Stalingrad in 1942 was a heavy blow for her. In all probability her consent to her transfer to Arzamas-16 stemmed from a wish to get away from her native city, where everything reminded her of her husband. She exchanged her uneventful job in a factory radiographic laboratory for the difficult and dangerous profession of explosives specialist. She measured the mass velocity of explosion products using flash radiographic methods. Petals of tungsten foil thirty microns thick were arranged on the charge axis near its base. These petals were subject to acceleration in line with the detonation of the charge, and by their displacement one could measure the mass velocity of explosion products.

The use of conical charges allowed for a substantial decrease in the mass of the explosive and, commensurably, in the discharge associated with the radial dispersion of explosion products. The value obtained by Sof'ina for mass velocity accorded well with the Landau-Staniukovich theory.

Sof'ina was a source of delight and amazement. She worked with explosives on our testing sites from morning till night. Here we see her placing a charge for radiographic experiments. She is

The ones who began

wearing a sheepskin coat. Everything around is covered in snow; only a narrow strip of asphalt has been cleared to allow access to the front wall of the bunker from the charge under study. I remember our conversation after she first told me of her proposal to use conical charges. I said to her: 'You know, Vera Viktorovna, your idea would do credit to any man.'

She took offence: 'What call is there for comparisons like that? Haven't you grasped the simple truth that, with rare exceptions, there's nothing a man can do that a woman can't? And when it comes to preparing experiments, a woman's fingers, slim and accustomed to work, are better than a man's.'

After her experiments with explosives, Sof'ina was one of the first in the country who had to turn to an entirely different area of experimental physics, one which bore absolutely no relation to X-rays—the study of the properties of the hydrides of such metals as zirconium, titanium, and palladium. She built the apparatus and devised the procedures whereby hydrogen interacted with these metals.

N.G. Pavlovskaia recalls:

> The first woman successfully to defend her doctoral dissertation at Arzamas-16 was Vera Sof'ina. She was then fifty-seven years old. The theme of the dissertation was the study of the hydrides of vanadium, hafnium, palladium, zirconium, and other metals. Her examiner, Professor D.A. Frank-Kamenetskii, was struck by the scope of her work. He thought that, before Sof'ina researched the subject, this information had been unavailable to anyone in the country. I remember the day she defended her dissertation very well. She was so nervous that she completely lost her voice. We had to take urgent action. In a flash someone came up with some fresh eggs, and, while the official documents were being read, she ducked down behind the rostrum and gulped them down. It clearly helped, since her voice sounded fine and she conducted her defence brilliantly. More than three decades have passed since then, but the research results obtained by Sof'ina have not lost their scientific and practical value to this day. They are still of use to scientists studying hydrogen-metal systems.
>
> Working with Vera Sof'ina was always interesting and entertaining, but not easy. Being extremely demanding towards herself and others, she put all her extraordinary research talent into the job in hand. She considered that the

103

most important thing was to immerse oneself in the problem, to be infected by it, to grasp its essence, and thereby achieve success. She had the ability to see a problem from all sides and isolate the critical elements. She was indefatigable in her research, was inventive, thought through everything carefully, and loved to come up with elegant solutions. With her thin, delicate fingers she would make models and mock-ups out of paper. Using plasticine and wire, she would make little crystalline grids of the metals with which she was working and use them for visual aids. She always tried to discuss with qualified experts the results she obtained and the questions they raised. Zel'dovich, Davidenko and Frank- Kamenetskii would often look in on our little laboratory. She would acquaint the head of her division with the results she had obtained practically every day. In these conversations Sof'ina not only knew how to listen but also how to defend her point of view stoutly, without deferring to authority. When she was right, she would celebrate a double victory, since she was convinced that women in science deserved equal status with men.

She had a profound sense of beauty. She loved nature. She liked going to see shows in our theatre. She had a large and skilfully selected collection of art reproductions and a great many art books. She knew and loved music, both serious and popular. We came to like much of the music she liked. Apart from her favourite *Introduction and rondo capriccioso* by Saint-Saens, she taught us to appreciate and understand Musorgskii, especially his *Khovanshchina.*

Sof'ina worked at Arzamas-16 for almost twenty years. She is one of the few women scientists to have been awarded a State prize and other governmental awards.

Diodor Mikhailovich Tarasov

HE was one of the first researchers at Arzamas-16. When discussions were being held about the tasks facing us, it became clear that first and foremost we needed specialists in technical radiography, in the physics of X-rays and in the physics of metals. Diodor Mikhailovich was just such a specialist. He arrived in Moscow in November 1946 and immediately became part of our research team. He had graduated from the physics department of Sverdlovsk University in 1937 and begun postgraduate work in

1938. When war broke out he had practically completed his doctoral dissertation on phased magnetic transformations of metal. In 1942 he was called up into the army and was not able to defend his dissertation until the very end of the war.

Tarasov was a tall, handsome man, more like an actor than a scientist. He was extremely conscientious, highly responsible, bold, demanding of himself and others and, at the same time, sensitive and kind. He arrived in Arzamas-16 in early April 1947, together with his large family. During his six months in Moscow he had fully mastered the techniques of researching shock and detonation waves. Apart from our flash radiographic units, all that we had available at the time were photochronographs with a sweep speed of up to a hundred metres per second. High-speed oscillographs and photchronographs were only in the developmental stage.

Tarasov was the first to begin conducting experimental explosions at our forest test sites. For these experiments steel barrels about two metres in diameter and four metres in length were made. The barrels were externally reinforced with concrete. These concrete and steel structures could withstand the explosions of charges weighing up to 1·5 kilogrammes.

Alongside his daily experimental work, Tarasov was active in public affairs. A number of our researchers at that time had not completed their higher education. The war had interrupted their studies. On Khariton's initiative a branch of the Moscow Engineering and Physics Institute was set up in Arzamas-16. The young researchers went there to complete their education. Lectures and seminars were given by Zel'dovich, Frank-Kamenetskii, Flerov, Zababakhin, and others. Tarasov was the first director of this evening institute, working on a voluntary basis.

In 1948 we had several flash radiographic units working on the forest test site for research into detonation and explosion phenomena. The increase in the number of these units and in the mass of radiographed charges called for a rearrangement of our leading scientific personnel. Thus Tarasov's group became an independent department. Their main task was to research into materials compressibility, using radiographic methods.

By August 1949 their research had produced values for compressibility which confirmed our theoreticians' predictions. In 1963, the year he defended his doctoral dissertation, Tarasov was awarded the Order of Lenin and a State Prize for his work.

Tarasov worked at Arzamas-16 for twenty-six years. There are at least three respects in which the adjective 'first' can be applied to him: he was our first researcher; he was the first director of experimental radiographic explosions at our test sites and he was the first director and organiser of the Branch of the Moscow Engineering and Physics Evening Institute.

After a heart attack in 1973, Tarasov retired. He died in December 1974.

PAVEL MIKHAILOVICH TOCHILOVSKII

PAVEL Mikhailovich Tochilovskii, one of the first laboratory technicians at Arzamas-16, began his working life in Moscow in 1946. Born in Odessa, he spoke in a mixture of Russian and Ukrainian. This strange language was mainly characterised by the use of phrases beginning with the word 'or'. 'Or don't I know where the Filatov Institute of Eye Diseases is in Odessa? Or don't I understand how to carry lead azide?' He spoke loudly and energetically. He joined in the laughter over his incorrect way of speaking. The smile never left his expressive face. He was tall and extremely animated.

During the difficult period of organizing and setting up Arzamas-16, Pavel Mikhailovich, or Pasha, as he was known, was indispensable. He could do anything: get a wedding present for a bride, get hold of a bouquet of flowers of incredible dimensions, take ten or fifteen kilogrammes of potatoes to Moscow—during his first year rationing was still in force there. He could carry out more delicate assignments as well. There was one case where it was only thanks to him that the marriage of one young researcher took place. However, his main job was to ensure our supply of explosive charges and initiators. It was not until about a year after our move to Arzamas-16 that we began to produce our own explosives. During that time secondary explosive materials, including our beloved lead azide, were delivered by plane from Moscow. Tochilovskii had sole responsibility for the delivery of these dangerous cargoes. I remember one amusing instance when Pasha's resourcefulness and entrepreneurial skills were shown to the full. A plant, which had been preparing our order, was slow in handing over the goods. When the vehicle carrying the explosives left the plant, the plane was due to take off in five minutes and it would take at least thirty minutes to get to the

The ones who began

airport. Pasha stopped at a phone-box, called the airport and said something along these lines: 'This is Pavel Mikhailovich speaking. We're not going to be in time to deliver the special cargo. The plane must be held for about forty minutes.' It goes without saying that the official on duty obeyed Tochilovskii's 'instruction' as if it were an order from his namesake, Major-General Pavel Mikhailovich Zernov. The plane was held, and the delicate cargo was delivered safely and on time to Arzamas-16. When this small piece of forgery was discovered, they wanted to punish the 'usurper'. But he said: 'When you think about it, what offence have I committed? My name really is Pavel Mikhailovich. And the official on duty forgot to ask me my surname.'

Over the long length of time he worked with us—over forty years—Pasha mastered the most varied of skills: he was a demolition expert, an electrician, went on long expeditions, and was in charge of producing electro-vacuum apparatus. He did all this with enthusiasm, at the highest possible level, working on the principle 'I want nothing worse, I can do nothing better'. His conscientious work won him a medal in 1955.

MIKHAIL ALEKSEEVICH KANUNOV

EXPERIMENTERS would find life difficult without good technicians and glassblowers. In the old laboratories, the technicians who caught on instantly and without drawings to what the experimental physicists had in mind were known as 'fine mechanics'.[17] We had several such masters. Among them, mention should go first to Mikhail Alekseevich Kanunov. On more than one occasion his inventiveness was the key to the success of our developmental work.

He was thirty when he joined our department, a short man, a little ponderous, with a tanned, very Russian face. He reminded you of the famous Levsha from Leskov's story.[18] His small hands with their long delicate fingers could work miracles. Looking through a binocular lens, he could engrave all twenty-

---

[17]Russian: *fain-mekhanik*, a hybrid of English and Russian.

[18]A reference to N.S. Leskov's story 'Lefty' ('Levsha', 1881), the hero of which performs prodigious feats of intricate metalwork.

three Cyrillic letters of his first name, patronymic and surname on a metal plate measuring one square millimetre.

Before he was transferred to Arzamas-16, Kanunov had worked as a chief shop foreman, and it was not easy to persuade him to join our department. Pavel Mikhailovch Zernov helped us to do this. When I told him that a technician of Kanunov's calibre would speed up a good many projects, PMZ thought for a bit, then said: 'Bessarabenko (the director of mechanical production) isn't going to give him up easily. Maybe I'll try and frighten him into it.'

Zernov phoned Bessarabenko and said something along these lines: 'Aleksei Konstantinovich! Tsukerman has an urgent job on here. He needs five highly qualified instrument makers.' After a short pause, Bessarabenko's agitated voice came over the speaker, which Zernov had connected to the phone circuit: 'Pavel Mikhailovich, you can't do that! We've got all sorts of work on. If you deprive us of five instrument makers, we won't be able to fulfil the plan.' 'All right, then,' replied Zernov. 'Have it your own way. Give Tsukerman three people.' 'That's still too many. You'll cripple the workshop.' 'OK. Come to my office at nine o'clock tomorrow morning and we'll discuss the staffing situation.' The next day, at nine-thirty, I met Bessarabenko as he was leaving Zernov's office. He was pleased. They had settled on transferring just one man—Kanunov.

There was not a single project of any significance at all in which Kanunov did not participate to a greater or lesser degree.

When we urgently needed special electrodetonators, Kanunov, with his superb grasp of instrument technology, managed to plan, develop, and produce the first batch within days. He gave them the abbreviated designation KM-2. I asked him whether those letters stood for 'Bridge Capsule'.[19] 'Not only that,' he answered. 'They're also my initials—Kanunov Mikhail.' In order to put these detonators into industrial production many thousands of them had to be tested. The plant, citing production overload, refused to deliver these units. At that point Kanunov suggested that a price be set for his detonators, and he would try to produce the components himself as quickly as possible. I passed the idea on to Zernov. The price was set at 2 roubles 72 kopecks per batch. PMZ agreed to put Kanunov temporarily on a piece-

---

[19]Russian: *kapsiul' mostikovyi.*

The ones who began

work salary. In the first month Kanunov produced four thousand batches of components. Accordingly, Kanunov was entitled to 10,880 roubles, a sum far in excess of the monthly salary of his department chief, who held the degree of Doctor of Science. Naturally, the accounts department refused to pay up. Zernov himself came to investigate this sudden leap in productivity. It turned out that Kanunov had introduced dozens of improvements in the standard production process. Instead of machining each organic glass component on a lathe, he stamped them out. Using Kanunov's adaptations, it was possible to produce a thousand organic glass chips in a single working day. Electrodes were knurled instead of being engraved. Zernov familiarized himself thoroughly with Kanunov's inventions and improvements, picked up a phone, and rang up M.G. Kienia, the chief accountant: 'Mikhail Grigor'evich, pay Kanunov in full. He's got innovations and inventions coming out of his ears.'

Kanunov got his money. His proposals were later introduced in firms producing similar units. Hundreds of thousands of detonators were produced in this way.

Another episode from Kanunov's career as an inventor could be entitled: 'How to win a *Moskvich* car with a tram ticket'.[20] Those are the words I used to Kanunov when I learned that one of our inventions had resulted in savings of more than ten million roubles. His contribution to that invention consisted in new technology for producing components. Together with this, his clever adaptations allowed for radical innovations in their large-scale production. His bonus for that invention was 25,000 roubles. At the time that was the exact price of a *Moskvich* car. More than a quarter of a century has passed since then and throughout that time the *Moskvich* has been a fine servant to the Kanunov family.

During one of Kurchatov's visits to our laboratory an incident occurred which belongs to the realm of 'physicists' humour'. This amusing little story could be entitled: 'How the Great Craftsman plucked a number of hairs from the beard of the Great Academician Kurchatov'. It happened like this: when we were telling Kurchatov about a number of laboratory innovations, he was especially intrigued by a manipulator which enabled us to

---

[20] The point appears to be that Kanunov travelled to work every day on the tram, work which ultimately earned him a *Moskvich* car.

109

assemble pieces of apparatus in high vacuum. The meticulous Kurchatov insisted on seeing Kanunov's brainchild in action. We took him into the room where the manipulator was. Kanunov let air into it and opened it up. Kurchatov bent down to get a better look and his head was thus inside the manipulator. Kanunov was especially proud of the automated pincers with which you could grasp components and move them to where you needed them. Kanunov set the pincers in motion, failing to see that part of Kurchatov's beard was now between the moving jaws. When the pincers snapped shut, they pulled the hairs of Kurchatov's beard. Kurchatov jerked his head back in surprise. A number of hairs were left in the jaws of the pincer. Kanunov still preserves these precious 'relics' together with the large collection of other 'valuables' which he had accumulated in the course of a long life.

Among the many decorations he won during the Second World War there is one which Kanunov holds particularly dear— that of Distinguished Inventor of the Russian Soviet Federal Socialist Republic.

ALEKSANDR ALEKSEEVICH ZHURAVLEV AND IVAN IVANOVICH IGNAT'EV

TWO of our chief glassblowers—Ivan Ivanovich Ignat'ev and Aleksandr Alekseevich Zhuravlev—came to Arzamas-16 at almost the same time, in August 1948. Both were highly qualified. Ignat'ev came from a family of glassblowers in Klin. This town is not only known for its Tchaikovsky Museum, but is also the centre of Russian glassblowing. In particular, the famous Petushkov dynasty of Moscow glassblowers came originally from Klin.

Aleksandr Zhuravlev (universally known as Uncle Sasha) was also a glassblower of the highest calibre. No-one knows how many glass components, taps, pumps, joints, and 'doughnuts' for betatrons were made by the skilled hands of Ignat'ev and Uncle Sasha.

In the early days we worked without gas; we used petrol-driven generators. They would often flare up. Depending on the situation, either Uncle Sasha or Ignat'ev would be reprimanded, but they carried on with their work. Both men were awarded the Order of the Red Banner for their work.

Part 4

TITANS OF THE HUMAN SPIRIT

In this chapter we will discuss our scientific directors, leading theoreticians, and two superb organizers of scientific work. To people unfamiliar with the problems of nuclear science, the proud words used in the title of this chapter may seem immodest. However, it would be difficult to choose any other title for the talented people who gave their hearts and souls, and nearly the whole of their conscious lives, to solving these problems.

ACADEMICIAN IGOR VASIL'EVICH KURCHATOV

THIS is how Khariton's wife, Maria Nikolaevna, recalled her first meeting with Igor Vasil'evich Kurchatov:

> In 1942, after the epic of Sebastopol,[1] I saw Kurchatov with a beard. I asked him: 'Igor Vasil'evich, why have you got that pre-Petrine decoration on your face?'[2] He replied by jokingly declaiming two lines of a song which was popular at the time: 'First we drive out the Hun; we only shave when that's done'. But even after the Nazis had been driven out, there was still no time to shave. The beard suited this tall and dignified man very well. Soon people started calling him 'The

---

[1] The Crimean naval base of Sebastopol fell to the Germans on 4 July 1942 after a siege lasting almost eight months.
[2] Peter the Great (1672-1725) famously banned the wearing of beards by his courtiers, officials, and military.

Beard' or, on occasions, 'Prince Igor'.[3] There was something about him of the *bogatyr*.

Even in our Kazan days we had had to maintain close contact with Kurchatov. At that time our laboratory was engaged in research on explosion phenomena. We needed some high-voltage condensers. A number of excellent condensers turned up in the corridor of the main Kazan university building. It was not difficult to find the owner—they were Kurchatov's. I went to see him and told him about our requirements. He listened carefully, gave the matter a second's thought, then said: 'Take them. It doesn't look like there'll be any occasion to use that Cockcroft apparatus[4] before the war's over, and that's what these things were originally meant for.'

I remember very clearly the first seminars held in Laboratory No. 2 in Moscow, the name given in 1946 to what became the Institute of Atomic Energy. Staff from the Institute of Chemical Physics, from FIAN, and a few other organisations took part in these seminars. Usually about twenty people would gather in an empty room, each person bringing a chair. The chairs were a motley lot, but Kurchatov was always given an old-fashioned upholstered armchair of unknown provenance. It had carved armrests and feet, and a high back. The seat and back were upholstered with bright-green plush.

He would arrive after a sleepless night, fresh from the shower, his hair still wet. He would listen almost without interrupting, without intervening, although, as a rule, we would talk about things which interested him.

When he chaired meetings, they would be lively and heated. He managed to ensure that everyone voiced their opinion in a clearly formulated way. He would ask everyone in turn for their opinion and, if he liked the answer, there would ensue the inimitable Kurchatov 'Quite right. Quite right'—with the emphasis on the drawled and rolled 'r'. 'Now Iasha Zel'dovich

---

[3] A reference to the opera by A.P. Borodin. There is a double reference to Kurchatov here. Borodin, which derives from the Russian word for 'beard' (*boroda*) was, according to Sakharov, a code name for Kurchatov. Igor was Kurchatov's first name.

[4] J.D. Cockcroft, together with E.T. Walton, developed the first proton accelerator in 1932.

wants to say something.' It seems that even now I can hear his powerful voice.

He was very fond of making up witticisms himself. During one session we were discussing a technical project which was going to require the participation of industry and considerable financial expenditure. 'Let's ring the adminiboys right now,' he said, dialling a number. I leant across to my neighbour and asked quietly: 'What are the adminiboys?' 'It's his abbreviation—administration boys.[5] It's the name he gives to everyone, starting with the deputy minister.'

We met again in Moscow at the beginning of 1947. I had come to Laboratory No. 2 to take delivery of some imported equipment. Kurchatov met me in the entrance-hall and led me to his office. We passed through a number of militarized checkpoints. The greetings of the soldiers and officers: 'Welcome, Comrade Academician' sounded odd. After some brief negotiations I persuaded Kurchatov to give me two oscillographs and four avometers. Kurchatov questioned me about the new methods we were employing in our research into explosion and detonation processes.

One of his most remarkable characteristics was his extremely respectful attitude towards people and their daily needs. June 1953 was hot. Together with a group of physicists I was working on the calibration of some measuring apparatus, one part of which was set up on a steel workbench, the other on some asphalt near the workshop. Nothing went right all day. The measurements we were getting from our test runs were significantly inconsistent. Kurchatov looked in on us every two or three hours. It grew dark. Suddenly, as if someone had waved a magic wand, we started reproducing measurements with great accuracy. By three o'clock in the morning we had tested a large proportion of the components. At four we got a phone call. Kurchatov wanted to know how things were going. We told him that, for some unknown reason, since nightfall the measurements had gone without a hitch. 'Not cold, are you?' 'Not too bad. We'll survive—it's almost morning.' A quarter of an hour later a car drove up. Kurchatov had sent us four sheepskin coats.

---

[5]Russian: *rukrebiata*, derived from *rukovodiashchie rebiata*. Kurchatov is parodying a type of word-formation common in Soviet bureaucratic language.

In the morning our measurements again went awry. Suddenly it dawned on us: the asphalt was melting in the June sunshine and the legs of the steel workbench were sinking into it. That in turn was altering the relative orientation of the components. Testing this hypothesis was simple: we set up both parts of the apparatus on a single steel plate. The 'solar effect' immediately disappeared, and we completed our measurements successfully.

Academician Gersh Itskovich Budker, an expert in accelerators, nuclear reactors, and plasma physics, told me the story of one incident which showed how Kurchatov conducted himself during the difficult and complex time of the 'struggle against the cosmopolitans'.[6] The leading player in this story was Budker himself. During a routine survey of questionnaires filed by staff, the representative of the personnel department decided he did not like Budker's surname. The appropriate arms of the security services suggested to Kurchatov that he deprive Budker of his security clearance and dismiss him from Arzamas-16. The situation was complicated by the fact that by this time Budker had been nominated as director of the Siberian Division of the Academy of Sciences Institute of Nuclear Physics.

The problem was solved in the following way: Kurchatov called Budker into his office and suggested that he limit his visits to Arzamas-16 to the library for a period of three to six months. Many months went by, during the course of which Budker meticulously visited the library. Periodically he would ask Kurchatov how long this voluntary abdication from his work might go on. Kurchatov took refuge in Ukrainian sayings such as 'Don't rush into the fray before your father'. This game went on until the Twentieth Congress of the Communist Party in 1956, when many values were re-examined.[7]

A particularly memorable meeting took place in a casemate on one of our test sites in the forest. Khariton had phoned to warn us that he was bringing Kurchatov with him to show him how we were conducting experimental radiographic research in explosion phenomena. In the reinforced-concrete bunker which housed the flash-radiographic apparatus there were three people apart from myself. Among them was an inveterate smoker,

---

[6]This was a euphemism for the anti-semitic campaign launched by Stalin in 1948.
[7]This is a somewhat guarded reference to the de-Stalinization programme launched by N.S. Khrushchev in 1956.

Tat'iana Vasil'evna Zakharova, a senior researcher. When Kurchatov and company arrived, Zakharova, in her nervousness, started smoking cigarette after cigarette, flicking the ash so it landed next to an explosive charge. I explained in detail how the charge was primed and handed one of our detonators to Kurchatov so that he could examine it in detail. Khariton snatched this dangerous object out of Kurchatov's hands. A successful experimental explosion followed. Fifteen minutes after our guests had left, the phone rang. It was Khariton. 'A car's been sent for you. Please come and see me at once.' I set off to face the music. 'How could you give Kurchatov a detonator? Don't you understand what he means for us and how he has to be protected from the slightest risks? And that Zakharova of yours! What on earth possessed her to drop burning ash next to explosives? I ought to reprimand you for violating safety regulations.'

I had many occasions during our meetings to observe this protective attitude towards Kurchatov, this striving to shield him from chance occurrences. It was prompted not only by an awareness of the great responsibility he bore on his shoulders, but also by the profound respect for him felt by everyone who worked with him.

Khariton himself testifies to one very important aspect of Kurchatov's character—his gift for influencing those with whom he came into contact:

> He had to enlist the participation of major specialists from many fields of science and technology. He had to tear them away from their various chosen subjects, in which they were often the acknowledged leaders in our country, sometimes in the whole world. Only Kurchatov could have done it. The brilliant metallurgist Andrei Anatol'evich Bochvar was especially hard to win over. Once I was one of seven people trying to persuade him, but Bochvar would not leave his beloved light alloys for all the world. I don't know whether the method of collective persuasion was often used, but on this occasion it bore fruit. About two hours into the procedure Bochvar capitulated.'[8]

---

[8]From Khariton's remarks at the session of the Scientific Council of the Kurchatov Institute of Atomic Energy, held in 1988 to mark the eighty-fifth anniversary of Kurchatov's birth.

Kurchatov visited our laboratory for the last time in November 1957. When he found out that we were researching mechanisms of vacuum breakdown, he asked us to show him a film we had made which captured the growth and development of micropoints on the surface of an electrode in high vacuum under electric field. At the very beginning of his work in the Institute of Physics and Technology he had studied scenarios of electrical puncturing of thin-layered insulation. Although thirty years had passed since then, he retained an interest in questions of the strength of electrical insulation.

That same evening Khariton invited Kurchatov and a number of leading scientists to his home. On a large table-tennis table supper for twenty people was organized. The atmosphere was cheerful and relaxed. There was dancing. For the most part people amused themselves with water-pistols, which someone had brought back from Lebanon, and with a little portable movie camera. We managed to get some brief footage of Kurchatov getting the hang of shooting with a water-pistol. The best pictures were of Kurchatov, Zel'dovich and Zababakhin enthusiastically squirting water over one another.

Academician Anatolii Petrovich Aleksandrov relates the story of his decades of work with Kurchatov:[9]

> Our first years of work on the atomic problem were marked with Kurchatov's burning enthusiasm. It infected everyone who was drawn into the hectic whirl of his activity. I worked with Kurchatov in various situations over the course of thirty years, and his extraordinary qualities as a human being, a scientist, and an organiser were always evident and always left a deep impression in the hearts of those who worked with him. I was always struck by his extremely high sense of responsibility for whatever he was working on, irrespective of its scale. After all, we take a casual, unserious, cavalier attitude to things which we consider unimportant. It was not at all like that with Kurchatov. I remember when we were in the physics technical college, all of us still under thirty, the worst thing that could happen was when Kurchatov was elected to the Local Committee. He immediately dug up some old obli-

---

[9]Transcribed from a tape-recording of A.P. Aleksandrov's remarks at a session of the Scientific Council held in 1970 to mark the tenth anniversary of Kurchatov's death.

gations which we had generously agreed to on various occasions, and then immediately forgotten about, and nagged us until we fulfilled them. At the same time there was nothing of the pedant in him. He would act with such joy and conviction that in the end we all got caught up in his energetic style, and did whatever we had promised to do.

We had already nicknamed him 'the General', but not in any bad sense. We called him 'the General' because he knew how to mobilize everyone for any activity which he considered crucial, and was a first-rate organizer of that activity.

When the war broke out, and my laboratory was working with the navy on protecting ships from magnetic mines, using a method which we had devised before the war, Kurchatov, in his desire to be of the greatest possible help at that difficult time, brought the whole of his own laboratory in on the work. Thanks to his organizing skills, his sense of responsibility, and his rare ability to interact effectively with people of the most varied personal qualities and from the most varied of backgrounds, our navy suffered minimal losses from magnetic mines. Special agencies were formed in good time, and both training of personnel and the production of the required technology were organized. Credit for this belongs in the main to Kurchatov, and serves to demonstrate his enormous talent.

As far as our current work is concerned, Kurchatov felt himself hugely responsible to his country. One can say that he delved so deeply into all the basic questions that it was only after the direction and organization of the work were totally clear that he moved on to other tasks.

In my view, our country was extremely fortunate that such an extraordinarily focused man was put in charge of our project. Kurchatov's personal qualities contributed enormously to the scope, tempo, and direction of the work, and to its eventual success. In those years there were no half-hearted participants in the work: everyone worked without sparing effort or time; the originator of this style of work was, of course, Igor Vasil'evich Kurchatov.

At the same time, however, Kurchatov was the liveliest of men, witty, cheerful, always ready for a joke. I made fun of his beard and presented him with a huge razor. He retaliated with a practical joke of his own: as I was leaving for a trip to the Urals, Kurchatov gave me a parcel for Boris Glebovich Muzrukov, who was at that time based in

Cheliabinsk. I arrived at Muzrukov's for Sunday lunch, bearing the parcel, just as Kurchatov had asked me to do. Muzrukov opened the parcel, burst out laughing, and said: 'This parcel's meant for you'. He gave it to me and said that Kurchatov insists that I make use of it straightaway. I opened the parcel—inside was a huge wig. Naturally I put it on and, I must say, it did wonders for me. I recently wore the wig with great success in a film, in which I played Fantomas.[10]

This was a happy time, because of the intensity of our work, because of the results we obtained, and because of the magnificent spirit of comradeship which Kurchatov created. I am fortunate to have worked with him for so many years.

The happiness of victory—a happiness multiplied by the importance of the work we were doing and in the enormous collective participation in its accomplishment—was frequently experienced by Kurchatov. He was a key figure in the discovery of nuclear isomers, the spontaneous fission of uranium atoms, and the commissioning of the first Soviet cyclotron in the Institute of Radiation in 1939. The blue-white phosphorescence of atoms ionized with fast protons was the harbinger of Soviet work on artificial radioactivity. On a cold December day in 1946, as the cadmium rods were withdrawn, the meters surrounding the little uranium assembly began clicking vigorously. That day is rightly considered to be the birth date of the first atomic reactor in Europe. However, perhaps the feeling of total victory did not come to him until that momentous morning in late summer 1949, when the bright flash of a nuclear explosion lit up the door of the reinforced bunker. Even before the arrival of the shockwave and the awesome roar of the nuclear explosion, Kurchatov's laconic 'It worked' signalled the end of the USA's monopoly on nuclear arms.

He also received enormous inner satisfaction in the summer of 1956 after the paper he gave at Harwell on the first Soviet work on controlled thermonuclear synthesis. The world's press reacted to this report given by a Russian professor with a flood of articles and publications almost as great as that which followed the first Soviet nuclear test.

---

[10] A detective in a series of French comic fantasy films of the 1960s, played by Jean Marais.

It is impossible to compare the level of these and other high points in Kurchatov's career. Forward, only forward: such was his motto. This courageous and charming man was the charismatic leader of an army of scientists and engineers and of dozens of research organizations and plants, taking them to the topmost peaks of science. Less than fifteen years separate the commissioning of the first atomic reactor and the end of Kurchatov's life, but everyone who worked alongside him or under his leadership will recall those years as the happiest they ever knew.

I would like to tell the story of Kurchatov's last hours from the words of Maria Nikolaevna Khariton. As we listened to her, it seemed that we, too, had been present at the 'Barvikha' sanatorium near Moscow on that fateful day, 7 February 1960.

Kurchatov had arrived on that Sunday morning to visit the Kharitons. He was in excellent spirits. After exchanging greetings, he paced back and forth across the room several times and, on seeing a radio in the corner, he pressed one of the knobs. The strains of an old waltz rang out. Kurchatov asked: 'Maria Nikolaevna, how many years would you say we've known each other?'

'About thirty years, Igor Vasil'evich. We lived in the same apartment building on Ol'ginskaia Street in Leningrad for ten years, from 1931 to 1941.'

'But when was the last time we waltzed together?'

'I honestly don't remember.'

'So let's dance then.'

Probably, the countdown, which Kurchatov had heard so many times during testing, had already begun: fifteen minutes left, fourteen... But no-one present could hear this dreadful countdown.

They danced a few *pas* round the table. Kurchatov escorted his partner to her seat, and said: 'Maria Nikolaevna, do you know, I very much enjoyed something the day before yesterday, Friday. I was driving down Herzen Street, when suddenly I saw a big placard near the Conservatoire. They were doing Mozart's Requiem. I hadn't heard that amazing music since Leningrad. I stopped the car and went up to the box office. There were, of course, no tickets left. So I went to see the manager.'

'You can't be serious. This concert's been sold out for the last three weeks.' But I got out my documents and persisted. In the end I managed to wangle a ticket in the sixth row. What

119

superhuman, unearthly music! The 'Lacrymosa' alone is incomparable. You wouldn't happen to have any records of it, would you?'

'Of course we do'.

'When you get back to Moscow, I'll make a point of phoning you for them. I'd love to hear the music again.'

Then, suddenly, and quite mischievously, he said:

'Maria Nikolaevna, please help me to solve a sexual question?'

'Which one, which one, Igor Vasil'evich?'

'I just can't decide which linoleum to choose for the floor of the new laboratory. I'd like to ask your advice...'

At this point he shook some linoleum squares of various colours out of his briefcase and onto the table.[11]

Kurchatov then put on his overcoat, took Khariton by the arm, and said: 'Let's go for a little walk, Iulii Borisovich, and talk some shop.' And the silent countdown continued. They went out into the park. It was a frosty, but sunny day. The bare branches of the trees were powdered with snow on top. Kurchatov chose a bench and brushed off the snow for himself and Khariton.

'Let's sit here for a while.'

Khariton started telling Kurchatov about the results of his latest research. Kurchatov, always very responsive in conversation, for some reason said nothing. Khariton was gripped by sudden alarm. He turned quickly to Kurchatov and saw that his eyes were glassing over. 'There's something wrong with Kurchatov,' he shouted, as loudly as he could. Secretaries and doctors came running, but it was too late. A small blood clot had occluded the coronary artery. The countdown had reached zero. Kurchatov was fifty-seven years old.

---

[11]This exchange revolves round an untranslatable pun. In Russian the adjective *polovoi* means both 'sexual' and 'pertaining to the floor'.

Academician Iulii Borisovich Khariton

> In all I do I feel an urge to get
> To the very essence.
> In my work, in seeking my path,
> In the bafflements of my heart.
> To the essence of past days,
> To their cause,
> To the foundation, to the roots,
> To the very core.
> *Boris Pasternak*[12]

IN recent years a great many books and articles have been published dealing with the life and works of Igor Vasil'evich Kurchatov. There is an absence of information on the scientific career and research work of Iulii Borisovich Khariton.[13]

Both Kurchatov and Khariton were guided by the same principles, such as an acute sense of responsibility for the task in hand, selfless dedication to science, spiritual purity, good will, and perseverance. Enormous mutual respect and trust formed the basis for their relationship. They met in 1925. In 1939, as young scientists, they became members of the Uranium Commission of the USSR Academy of Sciences, headed by Academician V.I. Vernadskii. In 1943, when Laboratory No. 2 was organized, Kurchatov brought in Khariton to work on the nuclear problem. From the very beginning, Kurchatov directed work on nuclear reactors and the production and enrichment of nuclear fuel. Problems concerned with the construction and functioning of the atomic bomb were delegated to Khariton. Over the course of the first thirteen years he was the chief engineer of our organization, and, from 1959, its scientific director.

The combination of these two equally remarkable scientists and organizers proved extremely successful. Outwardly they were very different: Kurchatov was a tall man of really heroic build, Khariton small, ascetically thin, and very active. One of my first

---

[12] The opening lines of the first poem in the collection *When the Weather Clears (Kogda razguliaetsia)*, published in 1957.

[13] When these words were written, this statement was true. Khariton's name is missing, for security reasons, even from the third edition (1970-81) of the *Great Soviet Encyclopaedia*. However, as will be seen from the bibliography, a good deal of material about him has been published since 1991.

meetings with Khariton took place in the corridor of a Moscow institute. Someone called out my name. I turned round and saw someone running towards me, almost a boy. Probably a student, I thought. It was Iulii Borisovich Khariton.

He was the youngest child of a Petersburg journalist, Boris Osipovich Khariton. His mother was Mira Iakovlevna Birens, an actress in the Moscow Arts Theatre company. In 1904, when Iulii Borisovich was born, his elder sister Lida was five and his other sister, Ania, three. Soon a governess appeared in the family. It was she who brought the children up and looked after them for twenty long years. An Estonian by nationality, she spoke perfect German, a gift which she passed on to the children in full measure by only using German with them. In 1915 the children, along with the governess, moved into three small rooms in the attic of a seven-storey house on Zhukovskii street.

There was a strange staircase in the house, the flights of which were partitioned off one from another. Anna Borisovna (Ania) recalls: 'In 1916 we had a fire in the house. One of the rooms somewhere on the second floor caught fire. Smoke was billowing into the attic, where we were. Twelve-year-old Liusia (as Iulii Borisovich was known in the family) did not panic but soaked some towels under the tap and gave them to me and Lida, so we could breathe through the wet cloth, then led us down the staircase, which was full of smoke, and into the yard outside.'

Books played an important role in Khariton's education. There were very many of them in the family library. To this day he remembers with special gratitude a ten-volume children's encyclopaedia, with, on the cover of each volume, a picture of a young woman telling a little boy and a little girl about the sciences. The books of the well-known popularizer of science Ia.I. Perel'man played no small role. In 1915 Boris Osipovich gave his son an elementary fixed-focus camera with 'dropping plates'. Khariton retained a lifelong enthusiasm for photography.

In 1916 Khariton entered the second year of a commercial school, which equated, in terms of the curriculum, to the fourth year of a modern school. It was here that the following episode took place. When Khariton came to class for the first time, the mathematics teacher decided to test the new pupil's knowledge of multiplication, and asked him: 'What's twelve times thirteen?' The reply came immediately' 'One hundred and fifty-six.' There were gasps of astonishment from the other children in the class, and somebody said: 'That settles it—Khariton's going to be first in maths.' At the school, as well as German, Khariton studied French. In summer 1917, after finishing his third year, he immediately passed his fourth-year exams and moved into the fifth year. Then he received permission to do the sixth-year course as an external student, and by summer 1919 he had earned his diploma certifying completion of all seven grades, having spent only three years in school.

From the age of thirteen Khariton began working. At first he worked in the library of the Writers' House[14] and then, after he had finished his schooling, he worked as an electrician in telegraph repair shops on the Moscow-Rybinsk railway. In 1920 he became a student in the electromechanics department of the Petrograd Polytechnical Institute. There he had the good fortune to be in the group taught by Abram Fedorovich Ioffe for physics. After a few lectures from that remarkable teacher and scientist Khariton realised that physics was a more interesting and wide-ranging science than electrical engineering. At the beginning of 1921 Khariton transferred to the department of physics and mechanical engineering, run by Ioffe. His fate was thus linked with Academician Ioffe for many years and with physics for the rest of his life.

In the same year, 1921, Nikolai Nikolaevich Semenov invited Khariton and two of his friends, A.F. Val'ter and V.N. Kondrat'ev to work in the laboratory which Semenov had recently set up in the Leningrad Institute of Physics and Technology. This is how Semenov tells the story of that laboratory's early years:[15]

> The Civil War had just ended. The country was racked by hunger and ruin. There was no equipment, no instruments. You would think that in normal times no-one could work in such conditions. But all difficulties were overcome through a mixture of enthusiasm, stubbornness and, I would say, optimism. Since the Institute of Physics and Technology was still under construction, the laboratory was housed in the Polytechnic Institute. In a freezing building, where the temperature in the corridors was often lower than it was outside, in a tightly sealed laboratory, everything was done by

---

[14]Russian: *Dom literatorov.* A short-lived (1918-22) organization established in Petrograd for the welfare of writers and the publication of their works. Not to be confused with similarly-named organizations set up in all major cities of the USSR in 1934.
[15]From an address given by N.N. Semenov on 27 February 1964 to a meeting of the Scientific Council held to mark Khariton's sixtieth birthday.

hand by Khariton and his two friends. They had a water conduit assembled as follows: on a high wooden platform stood a large ebonite vat, made from a submarine battery and filled with about twenty buckets of water. They carried the water there from a standpipe in the street or from neighbouring buildings. Pipes led from the vats to cool the diffusion pumps for all the refrigerators and other equipment which needed a water supply. There was a little stove which had to be stoked everyday, and finding firewood was not easy. From 1923 onwards, small amounts of imported equipment began arriving in the laboratory, followed by Soviet-made equipment. Food was often scarce and seemed to consist entirely of millet *kasha*. Money was measured in millions, and then billions, of roubles, but even so there was nothing to buy with it. To build the equipment we had one technician and one glassblower. Many pieces of apparatus we built ourselves from scratch. Nevertheless, some first-rate research was carried out which was published and became part of the world body of scientific knowledge. At this time Khariton did some outstanding work on the condensation of molecular bundles.

I would like to talk about the research mentioned by Academician Semenov in more detail. In order to study the interaction of a bundle of cadmium or zinc atoms with the surface of solid bodies at various temperatures, Khariton used the following elegant method. He dipped one end of a copper wafer 1 cm. wide by 13-15 cm. long into a cylindrical vessel filled with liquid mercury. A small heating element was wound round the other end of the wafer. The vessel was immersed in a Dewar of fluid air. The mercury would freeze with the copper wafer firmly inside it. Once the heating element was turned on, a constant temperature gradient would be established along the wafer from $+10°C$. to $-140°C$. A source of atom bombardment was set up opposite the wafer in the form of nichrome wire, on which the metal under study was deposited. When the wire was heated, atoms of the metal would fly off in the direction of the copper wafer. They would be reflected off that portion of the wafer whose temperature was higher than critical, and settle on the remaining cooler portion. The boundary between these two areas was quite distinct. Semenov's and Khariton's findings were refined by Knudsen-Wood's data.

These experiments revealed another unexpected fact. Although the temperature of the wafer, diminishing from the heating element towards the frozen mercury, was constant, the boundary between the areas of total reflection and total settlement of cadmium atoms or zinc atoms did not form a straight line, but, instead, described an arc. Ia.I. Frenkel, the director of the theoretical department of the Leningrad Institute of Physics and Technology not only calculated the form of the arc, but, on the basis of these experimental findings, devised a general theory of critical temperature of vapour settlement.

One of the most significant works of the young Khariton was done at the Leningrad Institute of Physics and Technology in 1925, together with Z. Val'ta. They researched phosphorus combustion in oxygen, work which became the point of departure for further studies of branched chemical chain reactions. This is what Khariton has to say on the subject:[16]

> In one of our discussions, my dear teacher Nikolai Nikolaevich Semenov and I concluded that it would be worthwhile to see if it would be possible to increase the specific light yield from the oxidation of phosphorous vapours by reducing the air pressure at which oxidation occurs...I rigged up an apparatus, and, following the usual practice for those times, got the more complicated components from the glassblower and soldered everything together myself. We set up a high vacuum in a vessel around a piece of white phosphorus, and admitted oxygen through a very fine capillary tube. The oxygen pressure in the vessel gradually increased, but there was no illumination at all. The phosphorous vapours refused to oxidize. And only when the oxygen pressure reached the hundredths of a millimetre of mercury did a steady glow flare up throughout the whole interior of the vessel. The oxygen pressure ceased rising. This continued until we turned off the oxygen supply to the vessel. As soon as the oxygen supply was turned on again, the reaction resumed instantly.

Later, it emerged that if a small quantity of noble gas was introduced into the vessel, then the illumination, for some strange reason, resumed under a smaller oxygen pressure.

---

[16]From Khariton's address upon being awarded the Lomonosov Gold Medal. *Vestnik AN SSSR*, 5, 1983, 56.

The research had to be terminated when Khariton was sent to Cambridge to study. The results of the experiments were published. A few months later, when he was already in Cambridge, Khariton came across an article by Max Bodenstein which was sharply critical of the results obtained by Khariton and Val'ta. In response, Khariton sent a letter to Semenov and Shal'nikov, outlining the need to refute Bodenstein's article. Semenov and Shal'nikov continued the work begun by Khariton. It turned out that the oxygen pressure at which the oxidization of the phosphorous vapours began also decreased when the dimensions of the vessel were increased. Semenov drew general conclusions from these results and used them as the basis for a theory of branched chemical chain reactions. Bodenstein was forced to concede defeat.

Iakov Zel'dovich later recalled:

> The chief interest of this early research was not the research itself, but rather the whole situation, on the cutting edge of science as it was at that time, as well as the enthusiasm and close working relations between the people involved. The senior member of that group, A.F. Ioffe, was barely forty years old. Semenov, who was considered a senior scientist, was thirty. Besides them, there were ten or more future Academicians and corresponding members of the Academy who were in their very early twenties. They worked, and exchanged views, with tremendous enthusiasm, oblivious of time. In that milieu was born the scientific school which we can now term the Ioffe school.

Khariton's visit to the Cavendish laboratories in Cambridge was an event so crucial to the development of Khariton the scientist that it merits special comment.

In 1921 Abram Fedorovich Ioffe received Rutherford's agreement for Petr Leonidovich Kapitsa to work in his laboratory. Kapitsa worked in Cambridge from 1921 to 1934. In 1926, on Kapitsa's recommendation, Rutherford agreed to take on yet another scientist from the Leningrad Institute of Physics and Technology—Iulii Borisovich Khariton. Semenov helped with the necessary paperwork, and Khariton left for a two-year stay in England. Khariton writes:

> The Cavendish Laboratories were built more than a hundred years ago, in 1874. 'Externally the building is Gothic in style. There is no proper main entrance. The entrance is via a small

staircase on the left, under an archway. At one end of the courtyard was Kapitsa's laboratory. In the workrooms the walls were simply of neatly-laid brick, with no plastering. This did not tally with one's usual conception of laboratory facilities, but was no hindrance to work. To the left of the arch, at the very top, was a room known as 'the garret'. Every new arrival was obliged to pass through the garret, working there for a certain time in order to master elementary working practices for dealing with radioactive substances. I, too, spent about a month there.

There was a small workshop in which worked a number of technicians and metalworkers. In exceptional cases, where an especially complicated piece of apparatus, such as a Wilson chamber, was needed, Rutherford would authorize an order to an outside firm. I had some skills with machine tools and the chief technician was therefore well disposed towards me. He would occasionally let me cobble something together myself, thus avoiding distracting the other overworked technicians. The machine at which I worked had no motor; instead it had a treadle, like a sewing machine. In Russia at that time such machines were no longer in use, but the English cling to their traditions. The machine had probably been acquired in the last century, but the technicians still used it.

Some remarkable work was done with this simple equipment. The use of meters and electronic amplifying lamps had scarcely even begun. The basic measuring instruments were still ionization chambers.

During my time in Cambridge I had occasion to meet many remarkable physicists. Paul Dirac, F.W. Aston, and the relatively young James Chadwick, were working at the Cavendish Laboratories.

The fact that it was Chadwick in particular who discovered the neutron cannot be ascribed to pure chance. After Rutherford's famous Bakerian lecture, given in 1920 to the Royal Society, Chadwick latched onto Rutherford's thesis that the atomic nucleus might contain particles similar to protons, but without charge. Rutherford postulated that there were protons in atomic nuclei so tightly bound up with electrons that they formed a sort of single, and neutral, particle. The idea inspired Chadwick, and for twelve years he laboured doggedly, searching for experimental proof for the existence of such particles. He searched for them both in the experimental work done at the Cavendish laboratories, where he had worked since 1919, and also in the results obtained at other physics laboratories engaged in the study of atomic nuclei. When Irène Curie and her husband Frédéric Joliot published their strange findings concerning a mysterious emission following the exposure of beryllium to alpha particles, Chadwick at once realised what had occurred in those experiments. He approached Rutherford with the idea that the mysterious emission was in fact neutrons. Rutherford, however, an extremely cautious experimental scientist, judged this impossible, although it was he himself who had posited the idea of the neutron. It is striking that Chadwick was so alert to the appearance of any experimental facts that, only a few days after the publication of the French work, he was already investigating the energy of the nuclei of different masses thrown off by the mysterious emission of beryllium under alpha-particle bombardment. He demonstrated unequivocally that the emission was none other than fast neutrons. The consequences of that discovery are well known, not only among physicists, but throughout the whole world.

Soon after my arrival in Cambridge the Cavendish Laboratories celebrated the seventieth birthday of their former director, the celebrated physicist J.J. Thomson, who had proved experimentally the existence of the electron and had created, at the beginning of the century, one of the first models of the atom. A whip-round was organized and Thomson taken out to a restaurant. The speeches were made by two physicists who had known Thomson at the height of his career, and who were also brilliant speakers. One of them was the Frenchman Paul Langevin, famous for his work on magnetism. He recounted how, during the course of their very first meeting, he had begun telling Thomson about his plans for his work in Cambridge. Thomson listened to him in desultory fashion, then interrupted him and said: 'Do whatever you like, only, please, don't forget to turn off the gas when you leave the laboratory.' Independence was highly regarded at the Cavendish Laboratories.[17]

While working in the garret, Khariton saw that newcomers like him were observing the flashes of light which arose in zinc sulphide when struck with alpha-particles. Since the energy of a

---

[17] Use is made here, and elsewhere in this section, of Khariton's public lectures given between 1976 and 1983.

single particle is not great, but the resultant flash is clearly visible, it became evident to him that the human eye is extremely sensitive. When Khariton asked how much light was given off by each flash, no-one could give him an answer. Khariton felt that the possibility of observing separate alpha-particles was guaranteed by this extremely high degree of sensitivity. He therefore asked Chadwick, under whose supervision he was working, to give him permission to calculate the minimum amount of light energy which could be detected by the eye. Chadwick consented and suggested he do this work with a young Canadian who had just joined the laboratory. The work was completed in two years. Khariton demonstrated that the sensitivity threshold of the human eye is extremely low. It is sufficient for some 15 quanta to strike the pupil for an eye well adapted to darkness to detect the flash.

This work was also the subject of Khariton's doctoral dissertation. In England the process of awarding academic degrees is straightforward. Khariton had to offer only one subject—physics. His examiners were Rutherford himself and Chadwick. All the rest took place in the spirit of mediaeval traditions. In one of the assembly halls of the university, before a considerable audience, Khariton and other candidates went down on one knee. A senior member of the university placed the fingers of the right hand in a particular manner on the head of each candidate, and said a few words in Latin. Khariton was then given the degree of Doctor of Philosophy. He brought his black gown and traditional four-cornered cap, the traditional symbols of British universities, back with him to Leningrad.

Returning to the Soviet Union via Germany in 1928, Khariton was alarmed to observe the growing strength of fascism, although German physicists of the time did not attribute much significance to this. When he got back to Leningrad, the twenty-four-year-old Khariton soon organized an explosive substances laboratory in the Institute of Chemical Physics, which had split off from the Institute of Physics and Technology. During the next twelve years he was to do a great deal of work on various aspects of the physics of explosions and detonation. Together with A.F. Beliaev, Khariton discovered the phenomenon of detonation transfer in vacuum between milligramme-scale charges of lead azide positioned some tens of centimetres apart. Their research demonstrated that detonation transfer occurs because of microscopic particles of admixture in the lead azide, which disperse at kilometric speeds. The weight of these particles is of the order of 10^{-11}g.

The research done by Khariton in 1939 with his young colleague V.S. Rozing played a special role in the understanding of the physical processes involved in the detonation of cylindrical explosive charges. It turned out that every explosive charge has its own critical diameter. If its value is less than a certain magnitude, the substance will not detonate; if it is greater, the substance will detonate. It was demonstrated that when the period of explosive product dispersion is less than the period of chemical reaction completion at the detonation wave front, the detonation attenuates.

In March 1944 Khariton reported on this research at a seminar chaired by P.L. Kapitsa at the Institute of Physical Problems in Moscow. Many physicists were present: Ioffe, Zel'dovich, Landau, Obreimov, Shal'nikov, and others. At the end of the seminar Kapitsa asked Khariton: 'Iulii Borisovich, why exactly have you told us this?' 'This is the reason why. Imagine for a moment that some Swiftian Lilliputians use as weapons grenades scaled down proportionately. If they arm these grenades with TNT, whose critical diameter exceeds 10 mm., they would not explode. RDX, on the other hand, whose critical diameter is less than one millimetre, would work extremely well.' After a short pause, he went on: 'I can give you another example. In Swift's work, as you know, we find giants as well as Lilliputians. This illustration has to do with large quantities of chemical substances, such as giants would have to deal with. A disaster occurred in 1921 at a chemical factory in Germany. During the First World War many thousands of tons of nitrate and ammonium sulphate had been stored at the factory. These compounds were used as fertilizers and were never considered as explosive substances. As it settled, the mixture turned into a large, hard, monolithic mass, which had to be made friable in order to put it into sacks for sale. And so they drilled into the consolidated mass and set small explosive charges. The heat of decomposition of the mound was four times less than that for TNT, and no-one foresaw any danger. At first everything went well. Many hundreds of explosions were carried out. Finally, however, when around 4500 tons of the consolidated compound remained, the whole mass detonated. The factory was destroyed. Many people died and there was damage to the town. If the mechanism of detonation and the dispersive sensitivity of chemical compounds had been better understood at the time, the catastrophe could have been averted.'

Let us return to Khariton's work during the 1930s. The classic work of this period was the theory of the separation of gases under centrifugation.

After the report on uranium fission given by Lise Meitner and Otto Frisch in January 1939, Khariton and Zel'dovich studied the possibility of a branched nuclear chain reaction in uranium. Khariton recounts the prehistory of the discovery of that phenomenon.

> It was like this. Shortly after the discovery of the neutron in 1932, the interaction of neutrons with various substances became widely used. The husband-and-wife team of Joliot and Curie discovered artificial radioactivity. Everything seemed to be in order: there were many interesting, useful, and entirely comprehensible results. However, under irradiation, for instance, of uranium, some strange phenomena would arise. Radioactive elements seemed to be forming whose position in the Periodic Table was too far away from that of the irradiated material. This was impossible—only a single neutron was being added, or, in the most extreme case, a proton was being knocked out. This meant that only closely neighbouring elements could be formed.
>
> And then an article appeared in one of the chemistry journals by the superb chemist Ida Noddack. While still a young woman, together with her husband, also a chemist, she had filled in one of the empty squares in Mendeleev's Periodic Table by discovering a new element, rhenium, named in honour of the River Rhine, which ran near her birthplace. Noddack had occasionally been to chemistry conferences in the USSR. In her article, published in 1934,[18] she discussed results from a number of papers on radioactivity. On the subject of unaccountable phenomena there was only a single paragraph—but what a paragraph! 'Could not one hypothesize,' wrote Noddak, 'that atomic nuclei might not only emit alpha-particles, i.e. helium nuclei, but that they might also split into two or three parts?'
>
> Physicists do not read chemistry journals, and chemists could not appreciate the significance of Noddack's idea. For that matter she herself, apparently, did not consider the

---

[18] *Angewandte Chemie,* 37 (47), 1934.

possibility that fission would necessarily be accompanied by a gigantic release of energy. The journal with Ida Noddack's explosive paragraph quietly gathered dust on bookshelves. It was not until early 1939 that Otto Hahn finally became convinced that, during neutron bombardment of uranium, radioactive elements from the centre of the Periodic Table would be produced. Lise Meitner and her nephew Otto Frisch then thought the problem through, and immediately published in the journal *Nature* their conclusion that the absorption of a neutron by uranium would be accompanied by the fission of the nucleus into two unequal parts, with the release of enormous energy.

And so these two women, Lise Meitner and Ida Noddack, stood at the very cradle of atomic energy. Immediate use of their idea in Germany might have dramatically altered the course of historical events.

Khariton and Zel'dovich set forth the results of their calculations in six crucial articles, five of which were published shortly before the war. The sixth appeared only in 1983. These papers demonstrated that the necessary conditions for sustaining a nuclear chain reaction do not occur in natural uranium. Nor do these conditions occur in a mixture of uranium and ordinary water. Khariton and Zel'dovich were the first to point out the need to enrich uranium with its light, fissionable isotope. One had to learn to isolate isotopes. In this connection Khariton's 1937 paper on isotope isolation by centrifuge acquired special importance. The method was widely used in industry.

The critical mass at which chain reactions should occur in uranium-235 was calculated incorrectly by Khariton and Zel'dovich as 10 kg. The error arose not through any theoretical shortcomings, but through imprecise knowledge of the physical constants which enter the equation. Values for these constants were so imprecise at the time that analogous calculations done by the physicist Rudolf Peierls in 1940 gave a critical mass of one kilogramme. We remind the reader that the critical mass for uranium-235 is 50 kg.

For seventeen years, between 1929 and 1946, Khariton was deputy editor-in-chief of the *Journal of Experimental and Theoretical Physics*. He was well known among physicists in the Soviet Union for his objective and careful approach to the articles they submitted.

In 1943 the work which Khariton and Zel'dovich had begun in 1939 broadened its scope. Laboratory No. 2, headed by I.V. Kurchatov, was set up to ensure scientific direction for the whole research complex needed to create nuclear weapons. In 1946 a branch of this laboratory—KB-11—was established. It was subsequently renamed VNIIEF. Naturally, Khariton and Zel'dovich were transferred there.

On 2 May 1945 a group of physicists, mostly from the Institute of Physics and Technology, among them Khariton, L.A. Artsimovich, and I.K. Kikoin, set off for Berlin to study the work of German nuclear scientists.. The group was led by Avraamii Pavlovich Zaveniagin. All were in military uniform. Khariton had the rank of colonel. Khariton describes the trip thus:

> The majority of German scientists, among them the eminent theoreticians Otto Hahn and Werner Heisenberg, had been evacuated to West Germany, and were under American control. However, a few scientists had remained in Berlin, and they willingly talked with us. From these conversations, and from a detailed tour of inspection of Berlin institutes of physics and chemistry, we realised that German nuclear physics was not very advanced. Kikoin and I told Zaveniagin this, and said that we must try to find out whether the Germans had stockpiles of uranium. The basic raw material from which uranium was obtained at that time was uranium ore from the Belgian Congo. It was very probable that Belgian stocks of uranium had been removed by the Germans. Zaveniagin supported our idea and put a car at our disposal. From our conversations with the German scientists we also learned that there was a building in Berlin, the Rohstoffgesellschaft, or raw materials company, where a card index was kept of everything the Germans had plundered from the countries they had occupied. This building was next door to Hitler's Berlin residence. It was staffed mostly by women, real fascists all of them, who declined to answer our questions as to how to find the records which interested us in this gigantic catalogue. As a result of prolonged and excruciating attempts to find our way round the six-storey catalogue, we managed to determine that there was uranium oxide, but we could not establish where it was. We were helped by conversations with Germans more favourably disposed towards us. We managed to identify a number of towns with affiliated card indexes, which indicated the locations of warehouses. Next came trips to these towns. In the card index of one of these warehouses

we found an indication of a town to which some uranium oxide had been sent. Unfortunately, the amount turned out to be small, the bulk of it having been removed by the military and used in the repainting of buildings (uranium oxide is bright yellow in colour).

Eventually, we found out that a certain quantity of uranium had been sent to the warehouse of a tannery situated to the west of Berlin. We turned to the local military commander. When he heard the name of the town, he said that he thought the tannery was in the American zone. We decided to check what he had said. Fortunately, the tannery turned out to be in the Soviet zone, right on the border with the American zone.

It was a small town, which had probably sprung up round the tannery. The tannery was under the control of an antifascist group. It consisted of a number of workshops and warehouses, some of which were crammed with sheepskins. In one of the last warehouses we discovered a fair number of small wooden barrels. There was a scrap of cardboard on one of them with the inscription U_3O_8. We breathed a sigh of relief. We reported to Zaveniagin and arrangements were made for transporting the uranium oxide to the Soviet Union. The total quantity was about 130 tons.

At that time nothing was known about the location of deposits of uranium ore in the Soviet Union. Intensive geological surveys were set in motion. The first deposits were found in mountainous terrain in Central Asia. The ore had to be taken from there in sacks on the backs of donkeys.

Kurchatov told me at one point that the uranium found in Germany brought forward the commissioning of the first industrial reactor for obtaining plutonium by about a year.

Khariton was well aware of the scale of the work that had to be done to create the atomic bomb.

> It was clear, (he relates), that creating the bomb would require colossal pressures to squeeze out fissionable materials, and that pressures of the required magnitude could be created by large-scale explosions. Moscow was hardly a suitable place for work of this sort, and finding a suitable remote area reasonably close to Moscow was no simple task. We visited many factories which had been engaged in arms production during the war. Finally, after a long search, on 2 April 1946 Pavel Mikhailovich Zernov and I arrived in the small town of Sarov, where St Seraphim had once worshipped. Here there

was a small factory which, during the war, had produced munitions, including shells for 'Katiusha' rocket-launchers. All around were impenetrable woodlands. There was plenty of space and a lack of population, and we were thus able to carry out the necessary explosions.

Thus, as Fate and human designs would have it, this holy town became the birthplace of Soviet nuclear arms.

Khariton's personal qualities were to have a decisive effect on the ethos, management, and working relations which developed at Arzamas-16 from the very outset. His most important character trait was his understanding of, and support for, new and constructive proposals. As you left his office, you would feel as if you had sprouted wings: wings of hope, because you felt understood and supported; wings of faith, because if Iulii Borisovich thought the same as you did, he would put his trust in you and everything would work out. You would be struck by his persistence and his ability to get to the heart of the matter. Nothing was too trivial for him. Indeed, it often turned out that something you considered trivial proved decisive in the solution of a problem.

One memorable session of the Scientific and Technical Council lasted over six hours. A representative of one of the leading research institutes in Moscow was putting forward proposals which, it was hoped, would solve a complex problem. Large allocations of money were needed to carry out the necessary research. Khariton was chairing the meeting. All the members of the council were noticeably tired. Only Khariton was paying close attention, making occasional notes on a piece of paper in front of him. He asked questions and sought clarifications.

The hands of the large clock were approaching seven o'clock. Council members were asking: 'Isn't it time to finish? Let's adjourn the meeting till tomorrow.' At first the chairman pretended not to notice these suggestions. At last, at eight o'clock, he acceded, and said: 'You're right, it really is getting late. Let's postpone the rest of this discussion until tomorrow.'

The next day the session opened with a statement by Khariton: 'Last night I analysed your reasoning,' he began, addressing himself to yesterday's speaker, 'and I made a few estimates for myself. Things don't work out quite the way you said.' Formulae and figures appeared on the board. 'As you can see,' continued Khariton, 'with your assumptions we can't expect anything good at all.' The speaker was clearly taken aback by this. For a while he tried to object. But escaping Khariton's iron grasp is no easy matter. In the end, he had to concede defeat.

Khariton has the rare ability to seek out the weaknesses in the scientific or technical solution of a new problem. This is not groundless criticism, finding fault in everything. As a rule, his criticism is well meant. When discussing new proposals or improvements, he tries to collaborate with you in finding a way out of any difficulties. He is especially demanding where the precision and reliability of experimental results are concerned. He has no time for careless or shoddy work.

To rule out any doubts or hesitations during important experiments, especially ones fraught with potential health risks for the service personnel, Khariton required that detailed instructions be drawn up. Our work demanded care and responsibility. It was precisely for this reason that in the most difficult years we had practically no serious accidents.

However, there was one famous occasion when the principle 'doubt and verify' nearly proved fatal to Khariton himself. Until 1954, work with assemblies which were approaching critical was allowed only under the direct supervision of Khariton. Operators were strictly forbidden to approach. On that memorable summer's day it so happened that, while Khariton was inspecting an assembly which was by then almost critical, he found that a disc which moved along a thread was loose. As he examined the thread (it turned out that the threaded rod had been badly made), Khariton bent forward and, as he did so, failed to notice that his head was too close to the assembly. What happened next took only a few seconds. The radiation meters, which until then had been steadily and regularly registering the neutrons and gamma rays emitted by the assembly, speeded up tenfold the moment Khariton's head approached it. Instead of clicks, they produced a continuous noise. One of the staff members who was monitoring the instruments which measured the approach to critical state gave a warning shout. Khariton asked an operator what the apparatus hanging on the wall and meant to show the total radiation dose was indicating. Back came the embarrassed answer: 'The apparatus isn't working.'

After this incident, Khariton had daily blood tests and we watched with alarm as his white blood cell count grew day by day. Khariton plotted these results, using a statistical curve published by the Americans, when, in similar circumstances, an operator had died. Later, Khariton admitted: 'When I looked at the graph, I thought my number was up, too, but I took solace in the fact that I had no nausea at all.' Fate was much kinder to him than it had been to the American. After a few days his white blood cell count began to fall, and soon it returned to normal.

Khariton is possessed of the gift of attracting people. Many specialists who began working with him in the forties or fifties consider themselves extremely fortunate for having had the chance to work under such a director. In one of his last articles Zel'dovich wrote:[19] 'I consider it was a huge stroke of luck, an enormous piece of good fortune in my life to have been a friend and acquaintance of Iulii Borisovich for fifty years and, especially, to have worked under his directorship for twenty years.'

What is the secret of the special charm of his personality? A genuine human being and a genuine scientist is the sum total of many ethical and psychological categories. We will deal here with three of these, which seem especially pertinent: honour, humility, and kindness.

Some one hundred years ago Vladimir Dal defined 'honour' as 'the inner ethical worth of a human being'. For Khariton, the notion of the 'honour' of a scientist is sacred.

Nowadays you often hear the view expressed that if a piece of research is carried out in a laboratory headed by a certain person, that person has the right to add his name to the list of authors of the article, paper, or invention. Often it is not only the directors of laboratories, who think this way, but also, more lamentably, the scientific directors of whole institutions.

Concerning co-authorship Khariton holds to the same strict ethical standards as did Kurchatov, who famously declined to be co-author of an article on the spontaneous fission of uranium, the result of research conducted under his supervision by G.N. Flerov and K.A. Petrzhak. He said: 'If I put my name to that work, the honour of the discovery will be ascribed to me. One must not eclipse young scientists with one's own name. As for guidance and help—these are sacred obligations for a laboratory director.'

Over the past four decades Khariton has supervised an enormous number of research projects, but has only published one or two articles himself. He has been prevailed upon to put

---

[19] *Priroda*, 6, 1983, 102.

his own name to a paper only where his own contribution to the work's creative component proved decisive. And this despite the fact that he has personally presented dozens of his colleagues' papers to the journal *Papers of the USSR Academy of Sciences*, and that he devotes a great deal of time and effort to correcting and revising others' articles until they are finally suitable for publication. While dealing with one such article, destined for the journal *Nature*, he decided he did not like the English translation. In the course of a single evening he did a new translation. The editorial board of *Nature* corrected only one word in the entire text. When colleagues put pressure on him to accept co-author status, he invokes Kurchatov's arguments, then usually adds firmly: 'I would consider it dishonourable of me if I were to put my name to a piece of work for which I had done so little.'

What strikes one about his face? A high degree of intelligence, of course. But the main thing that strikes you about it is kindness. Iulii Borisovich is a very kind man, and always approachable.

Apart from his enormous contribution to the creation of the atomic bomb, Khariton was always very active in public affairs. From 1950 to 1989 he was a Deputy of the Supreme Soviet of the USSR. He took his duties very seriously and was genuinely upset if ever it proved impossible to render genuine assistance to those who elected him. No letter or request went unanswered, and a great many requests, on a whole variety of topics, came his way. In particular, Khariton managed to get gas laid on for a number of the regions for which he was elected.

On Khariton's desk there is a picture of a beautiful young woman. She is Maria Nikolaevna Khariton, née Zhukovskaia. The role she played in the accomplishment of the titanic task which Iulii Borisovich took upon himself is highly significant, and not immediately obvious to an outsider. In the 1920s she was an actress at the *Balaganchik* musical theatre in Leningrad. Khariton often went to this theatre and was especially enchanted by the young actress. Once, when he was visiting some physicist friends, Maria Nikolaevna unexpectedly showed up at their house. She spent the entire evening singing and dancing, and Khariton escorted her home.

They were soon married. What could they have had in common, an actress and an experimental physicist, both extraordinarily gifted, she boundlessly in love with her art, he totally absorbed by physics? Maria Nikolaevna abandoned her art, which she loved so dearly. She studied English, French, and German, taught Russian to foreigners, and ran classes for postgraduate students. From 1943 to 1947 she was on the editorial staff of a physics journal, translating articles from Russian to English. She did all this to create favourable conditions for her husband's work to develop. Her self-sacrificing love became a great source of energy for Khariton. She and her husband also shared a love of music, and the sound of concerti by Rachmaninov, Wagner, Grieg, Haydn, Chopin, and Sibelius could often be heard in their home. They were both well versed in prose, poetry, and painting. It was a great pleasure to be in their company.

Maria Nikolaevna was acquainted with many of the leading literary figures of the time, among them Maxim Gorky, Leonid Sobinov, Vladimir Maiakovskii, Evgenii Shvarts, Rina Zelenaia, and the well-known theatre directors Nikolai Akimov and Nikolai Petrov. She told wonderful stories, recreating scenes from the past and painting portraits of departed friends. She was an enchanting woman, with a rare gift for compassion and for loving others. People of all ages, young and old alike, were drawn to her, and treasured any chance to be in her company.

The lights stay on in Khariton's office until late at night. His day begins at eight o'clock sharp and goes on for at least fourteen hours, just as it did in the early days. He keeps to the same schedule at weekends and on public holidays. Even when he is ill, he does not stop working. When friends enquire about his health, he replies: 'Soon this meeting will be over and then I'll take my medicine.'

Detailed discussions go on about the problems which arise daily. Khariton disregards no-one's advice. In his office, alongside laboratory directors and professors, you can find ordinary technicians or skilled machinists, whose opinions Khariton often values as highly as those of factory managers.

Another brief story, told by Khariton himself, sheds light on his attitude to the natural world:

> This was in the fifties. At our pilot factory a technological procedure was being used which had been developed at an institute headed by Academician Andrei Anatol'evich Bochvar. Bochvar came to see how we were getting along with his technique. Everything went smoothly and the work was finished by the evening. It was the end of June, the time of white nights. The next day I had to be in Moscow. Suddenly I had the idea of inviting Bochvar to make the

exotic trip to Moscow during the white nights. I can remember it to this day. Not far from Nizhnii Novgorod[20] our driver saw something unusual ahead of us on the road. He switched off the engine and we coasted noiselessly up to an enormous badger. I never thought that badgers lived in such densely populated areas. Bochvar and I were both delighted with this unexpected find.

A passionate traveller, Khariton has been to many European countries—England, France, Germany, Holland, Poland, Czechoslovakia, and the USA. He has also travelled extensively within the USSR. In 1956 he and Kurchatov visited Central Asia and in 1980 he was in Kamchatka, admiring the Valley of the Geysers and other sights. In 1983 he spent a month on Sakhalin and visited the Kurile Islands of Kunashir and Shikotan. He often spends his brief summer vacations in a beautiful corner of Estonia, the little town of Ust'-Narva. His favourite pastime there is to take brisk strolls with his camera along the firm, sandy beach. He walks so briskly and tirelessly that younger friends can barely keep up with him.

'To be a scientist is a great happiness...' That is how Khariton began his concluding address at a session of the Scientific Council held in 1964 to mark his sixtieth birthday. 'There is no greater pleasure than the discovery of the new and the unknown, the opening up of paths that no-one before you has ever trodden. But to be a scientist is not only a great happiness, but also a great responsibility. We must remember our obligations and work for the benefit of ordinary people.'

ACADEMICIAN IAKOV BORISOVICH ZEL'DOVICH

HE handed me a grey tube about twelve millimetres in diameter and about fifty centimetres long, like a piece of spaghetti, and said: 'It would be very entertaining to see how fast this thing burns at the centre and at the edges. At the moment we're studying the combustion rate of powders under dozens of atmospheres of pressure. It seems to me that the central area of the tube ought to burn more slowly than the periphery. You've already taken

---

[20]See Part 1, page 11, note 7.

X-rays of bullets in flight. Can we not use your technique to study powder combustion phenomena?'

That's how I remember one of my first encounters with Iakov Borisovich Zel'dovich in January or February 1943. That was in the middle of the war, when fighting was under way for Stalingrad, and the great battle of Kursk was still only in the offing. At that time we had been successfully radiographing small-charge explosion phenomena. Zel'dovich followed the progress of this work closely.

Three years later we met again on another meridian, when Zel'dovich was appointed director of our theoretical department. Time demonstrated how right this decision was. A physicist with a wide background, with an excellent command of gas dynamics and the physics of explosions, Zel'dovich was the driving force behind our theoretical department for two years. He dealt superbly not only with the whole range of theoretical issues bearing on the weapons being developed, but also with the overwhelming majority of our experimental procedures, many of which were begun on his initiative. He would pose relevant and as yet unresolved questions to the experimenters, and often propose his own means and methods for resolving them.

Zel'dovich was instrumental in our acquiring yet another excellent theoretical physicist with tremendous inventive potential—Evgenii Ivanovich Zababakhin. After graduating from the Zhukovskii Air Force Academy in 1944, Zababakhin took up postgraduate studies there. His supervisor, Professor D.A. Ventsel, suggested 'The Study of Processes in a Convergent Detonation Wave' as his dissertation topic. When it was completed, the dissertation was sent to the USSR Academy of Sciences Institute of Chemical Physics. There it fell into the hands of Zel'dovich, who immediately saw that Zababakhin's work had a direct bearing on our own. In the spring of 1948 Captain E.I. Zababakhin joined the staff at Arzamas-16, where he soon became one of the most important members of the theoretical department. His contribution to applied nuclear physics and to the solution of a whole range of scientific problems connected with the creation of nuclear weapons was enormous. In 1968 he was elected a full member of the USSR Academy of Sciences.

Intuition, knowledge, talent...Without these three components no creative work in science is possible. Two decades of working with him convinced us that Zel'dovich possessed these qualities in full measure. They enabled him to seek and to find correct solutions to the most complex physical or physical-chemical problems.

With Khariton's permission, I'll give you an example. A meeting had, despite lengthy discussions, failed to come up with a correct solution. Then Zel'dovich appeared. He wrote a sequence of formulae on the blackboard, and within minutes, as if by magic, the problem was solved. When everyone had dispersed, and only Khariton and Kurchatov were left in the room, Kurchatov said: 'There's no getting away from it, Iashka[21] is a genius.' It is difficult to draw the borderline between talent and genius. Maybe there is no such border...

Once, after a vital experiment had been conducted, a large group of scientists was waiting for the oscillograms and data from other instruments to be recorded. Kurchatov said: 'It's a pity Iashka isn't here. He'd have whipped out his little slide-rule and given us all the figures we needed.' In those far-off days the slide-rule was the basic calculating instrument for both theoreticians and experimenters.

Zel'dovich was especially intolerant towards all forms of pseudo-science. At the beginning of the 1960s the so-called Din's Machine, which used gravitational forces, was the subject of much discussion. There is in our possession an open letter, dated 18 January 1963, which Zel'dovich addressed to experimenters:

> I have the greatest respect for our experimenters...They are an institute in their own right and have made an invaluable contribution to our common cause. I therefore find it not only funny, but painful also, to see an advertisement for Din's illiterate 'invention'. Anyone familiar with the fundamentals of mechanics can clearly see that it's rubbish. One can speculate as to why this rubbish has been patented or why it has ended up on the pages of a journal, but such speculation is not very interesting.
>
> Sometimes people say that, of course, Din's experiment contradicts theory, but nevertheless, it is an experiment and facts can be recalcitrant, and so on and so forth. However, people must understand that theory—in this instance classical mechanics—is the distillation of an enormous number of experiments which have been published, verified, widely discussed, and shown to be consistent one with another.'

---

[21]Iashka is a diminutive, affectionate form of Iakov.

Titans of the human spirit

Generations of physicists have been brought up, and will continue to be brought up, on Zel'dovich's superb books. One has to admire his ability to get to the very essence of complex problems, and his vivid, often figurative language. Many pages in his books are a veritable hymn to science and inspired creativity, and we well remember his survey articles in the journal *Achievements in Physics*, written in an uninhibited style, with a streak of mischievousness. Zel'dovich's oral presentations always evoked genuine pleasure and delight, now matter how diverse his audience may have been.

Zel'dovich was a man of all-round education. He often quoted poetry in his scientific articles. He defended his literary and artistic sympathies as energetically and consistently as his scientific convictions. He was vociferous and impassioned in his support for the election of the writer Chingiz Aitmatov to full membership of the USSR Academy of Sciences.

One of his English colleagues is reported to have told Zel'dovich, half-seriously, half in jest: 'I'm glad to have the opportunity to meet you and see for myself that you really do exist and are not just a collective pseudonym for a whole group of Soviet physicists.'[22] A three-time Hero of Socialist Labour, in 1956 Zel'dovich became one of the first to receive one of the newly revived Lenin Prizes for scientific research.[23] In 1958 he was elected to full membership of the Academy of Sciences.

When our work on this chapter was in full swing, a telephone call brought us the sad news that Zel'dovich had died suddenly from a heart-attack on 2 December 1987. Everyone who knew Zel'dovich was terribly shaken by his untimely death. The last time I had seen him was in November 1985, at the funeral of Khariton's daughter Tat'iana. I had suddenly felt someone staring at me, had turned round and seen Zel'dovich. We embraced warmly. Who would have guessed that this was to be our last meeting? He had never complained of heart trouble and had never been to the doctor.

At the civil funeral, held on 7 December, Andrei Dmitrievich Sakharov said: 'Iakov Borisovich Zel'dovich has passed away. It is very hard to believe this, since the idea of death does not

---

[22] The Englishman was the celebrated physicist Stephen Hawking.
[23] Lenin Prizes were not awarded between 1935 and 1956.

accord with our image of him. It is unbearably bitter to face the fact that he is no longer with us.'

What always struck one about Zel'dovich was his indefatigable scientific energy, his lively interest in everything new, his extraordinary versatility, and his intuition. He began working at an early age and worked till the very last day of his life. In that time he accomplished an incredible amount, in the most diverse fields. In the latest number of the journal *Achievements in Physics*, published after his death, there is an article of his which, as it were, throws a bridge back to his earlier work, when he specialized in chemical physics, surface phenomena, combustion and detonation, as well as chemical and nuclear chain reactions. Then he moved on, first to jet-propulsion technology, and then to the development of atomic and thermonuclear weapons. It can be said without any qualification that his work in this sphere was exceptional. He was responsible for a number of outstanding pieces of work on the physics of elementary particles. Here we can see the seeds of the future 'algebra of currents', a prediction of the existence and properties of the Z-bozon, and a statement of the problem of the cosmological constant. The last twenty-five years of his life were devoted to astrophysics and cosmology.

Zel'dovich's influence on his students and colleagues was striking. Thanks to him they often discovered within themselves capacities for creative work which might otherwise have gone partially, or even totally, unrealized. I had occasion to work alongside Iakov Borisovich for many years. When I reflect on that time, I realize how much I owe to his influence and ideas.

In science Zel'dovich was a man of enormous appetite—in the best sense of the word—and, at the same time, a man of absolute honesty, self-critical, willing to recognise when he was wrong and others were right and to acknowledge the authorship of other people. When he succeeded in doing something substantial or overcoming a methodological difficulty through some elegant manoeuvre, his joy was child-like. Setbacks and mistakes upset him deeply. He was very modest in his attitude towards science; he often felt that he was a dilettante, insufficiently qualified in this or that area, and made strenuous efforts (which an outsider could never detect) to overcome the gaps in his knowledge.

Our relationship had its difficult patches. There were bitter grievances and periods of coolness. Now all this seems mere foam on the stream of life, but, as they say, what happened, happened.

Academician Igor Evgen'evich Tamm

He did not so much enter the laboratory as run into it—a small, quick man, with kind, attentive eyes. He would greet everyone in turn with the words 'What's the latest?'

This was 1943, in wartime Kazan. The majority of the Academy of Sciences research institutes which had been evacuated from Moscow and Leningrad had set up operations in the grounds of Kazan State University.

We were X-raying explosion phenomena. Tamm was both interested and surprised by everything: the kenotron we were using for an X-ray tube with short-term cathode overheats and our means for synchronizing the X-ray flash with the desired phase of the explosion. Above all he was interested in the results we were getting from our radiographic experiments. He examined our first X-rays of lead azide explosions with great interest and took an active part in discussions about the new techniques.

My meetings with Tamm became more regular in the years 1949-55. At that time he was heading a group of young physicists who were conducting research into thermonuclear synthesis. He did not hesitate to cease working on fundamental problems of theoretical physics in order to take up the applied scientific work which was so important for national defence. In 1948-1949 our laboratory discovered the phenomenon of high electrical productivity in explosion products and dielectrics under powerful shock impressions. Resistance in explosion products near the detonation wave front turned out to be several orders lower than theoreticians had predicted. This was a question of fundamental significance. Tamm supported us actively. We have kept the favourable report on our work which he made in 1950.

Tamm's intolerance of any violations of scientific ethics was well known. Veterans of the theoretical department of FIAN affirm that throughout the whole time that Tamm was head of department, there were never any disputes about priorities. By 1970 there were more than forty highly qualified physicists on the staff. Each of them was an independent, creative individual. The microclimate created and maintained by Tamm practically eliminated the possibility of disagreements and dissension of the sort which are so detrimental to the creative atmosphere of even small teams of scientists.

Here is an instance when Igor Evgen'evich taught the author of these lines a lesson in ethics and behaviour. A stormy meeting

145

was in progress in Khariton's office. A large group of scientists had just been awarded state prizes and medals. Some key figures had been dropped from the list of laureates. I made an emotional speech, pointing out the injustice of the situation (I was not, of course, talking about myself). On the way home after the meeting Tamm said: 'In principle, you're right. Even so, you have to understand the purpose of any remark. Do you not think that your speech today was unethical in relation to Khariton for whom you have so much respect?' I have remembered this half-hour conversation ever since. Kindness, attention to the needs and cares of others, no matter what their social position, and a highly principled attitude were wonderfully combined in that man.

There are people who do good 'on credit', in the expectation of receiving some good in return. And there are also those who take satisfaction from an awareness of having done good and have no expectation of gratitude. Tamm belonged to the latter class of people—he did good quite naturally, without thought for the possible consequences to himself. His was a humanism founded on firm principles and profound convictions, a humanism that took no heed of temporary circumstances.

In the years when genetics was under attack in the Soviet Union,[24] Tamm enthusiastically greeted the discovery by Watson and Crick of the structure of the gene, the vehicle of heredity. In his brilliant lectures on genetics between 1956 and 1964 he popularized the significance of the discovery of the double helix and the genetic code. These lectures drew enormous audiences of people with the most varied of specializations. Under Tamm's influence Kurchatov organized a biology department at the Institute of Atomic Energy. Tamm worked closely with this department until 1967.

Here are two more examples which illustrate Tamm's principles. It was 1952. Tamm learned that a certain physicist, who considered himself one of his pupils, had joined the group of philosophers who were accusing Albert Einstein of Machism and

---

[24] A reference to the campaign of vilification headed by T.D. Lysenko, President of the Academy of Agricultural Sciences from 1938. In 1948 he claimed the authority of the Central Committee of the Communist Party for his campaign against the science of genetics. Lysenko retained Party support until as late as 1958.

idealism.[25] The man's article was read out in the flat of a close friend of mine. Greatly agitated, Tamm rose from the table and announced: 'You must understand that what is outrageous is not simply the fact that he sings the same song as those philosophers who don't understand the first thing about the theory of relativity. In the article he writes things he doesn't believe in himself.' They say that some time after this incident, there was an empty chair next to Tamm at some meeting or other. The physicist who had written the offending article came in late, sat down in the empty chair, and greeted Igor Evgen'evich. Instead of replying, Tamm rose demonstratively and took another seat further away.

During approximately the same period an unjust decision was taken to dismiss a certain mathematician from Arzamas-16. When this became known, some members of staff tried to ignore their colleague when they met him, although they knew he was not guilty of anything. By contrast, on the day of the man's departure, Tamm warned his colleagues: 'I won't be around after lunch today—I must support M.[26] and help him get ready for the journey.'

One of Tamm's students, Evgenii L'vovich Feinberg, writes as follows about his character and style of work:

> The concept of decency became particularly fully developed [...] among the best strata of the Russian working-class intelligentsia during the late nineteenth and early twentieth centuries. It was from there that it came down to us. In Igor Evgen'evich these traits were combined to a degree so extraordinary that it allows us to consider him as something of a role model.
>
> Perhaps one of the most important of these traits was his spiritual independence, in matters both great and small, in life as well as in science.
>
> Tamm had a profound sense of his own worth. I would even venture to say that he was a proud man. However, in using that word, I need to explain a great deal. His was not

---

[25]'Machism and idealism' was a smear frequently aimed at ideological opponents of Soviet Orthodoxy. Machism, the philosophy developed by Ernst Mach and Richard Avenarius (1843-1896), was a target of Lenin's anti-idealist critiques.

[26]It becomes clear from the section on A.D. Sakharov that the person referred to here is M.M. Agrest.

the pride that vulgar people equate with haughtiness. The Russian intelligentsia, from which Tamm came, had worked out its own specific standards.

He made great demands on himself and an enormous part of what he wrote was never reflected in publications. He published only things that would truly produce results, and the quantity of his published work is, by today's standards, unbelievably small, a mere fifty-five scientific articles, not counting popular articles, surveys, and reprints in foreign languages.

His tireless work, his enormous erudition, his ability to combine a physicist's approach, a physicist's understanding of the essence of a problem with a convincing mathematician's treatment of it, were remarkable examples for his pupils and colleagues.

Tamm's sixtieth birthday was marked in the summer of 1955. Khariton presented him with a large cake bearing the inscription 'I.E. Tamm—30 years', and said: 'Igor Evgen'evich! I knew, of course, that your birthday was coming up, but I had completely forgotten how old you were. I asked our theoreticians, who replied: "Hold on a minute. We'll do some calculations." They rang back to tell me that Igor Evgen'evich was thirty years old today. Later, it turned out that, as usual, the theoreticians had lost a factor of two, but by then it was too late to correct the mistake. The inscription had already been done. Besides, when I thought about it, I decided the inscription was entirely fitting. What do you mean, sixty? You don't even look thirty.'

Psychologists tell us that the ability to marvel is one of the main factors which determine a person's psychological and physical youth. Igor Evgen'evich retained that ability to the end of his life.

ACADEMICIAN ANDREI DMITRIEVICH SAKHAROV

> None is of freedom or of life deserving
> Unless he daily conquers it anew
> Goethe: *Faust* [27]

WHEN Andrei Dmitrievich joined us, it was immediately clear that a great talent had arrived, an absolutely extraordinary personality, with a mind that worked according to its own inimitable programme. Outwardly he was a phlegmatic young man, modestly, even carelessly dressed, the typical literary image of the scholar, cut off from the world and ordinary people.

This outward appearance was deceptive. Sakharov's range of social and academic interests was almost limitless. Just as limitless was his restless, indefatigable concern for the fate of other people, both his close friends, and others. To regard him as a theoretician, in the usual sense of the word, is quite wrong. Besides working on subtle theoretical problems, he, like Enrico Fermi, possessed a superb understanding of experimentation, and had an easy relationship with those who conducted them. He was a talented physicist, an inventor of genius, and an intuitive generator of ideas.

As often happens with gifted people, Sakharov was easy about his own ideas, did not insist upon them, and did not worry whether or not he was first in the field with them. As far as physics was concerned, his intuition was quite remarkable. He literally foresaw results through the most complex of formulae. I remember one such episode. Zel'dovich was giving a seminar about his new work on astrophysics. Complex theoretical constructs, weighty formulae, subtle evaluations. The audience listened intently; there were questions about his reasoning. Suddenly, Sakharov put a question. He could already see the expected result. Zel'dovich paused for a moment and, taking stock of the situation, turned to Sakharov and said: 'You ought to have gone for a walk for an hour, Andrei Dmitrievich. By then we might have had time to arrive at your solution.' Where did he get this ability to see the result? It was partly nature, partly nurture.

---

[27] This quotation, from *Faust*, Part 2, Act 5, is also used by Sakharov as the epigraph for his essay 'Reflections on Progress, Peaceful Coexistence, and Intellectual Freedom' (1968). The translation given here is from Philip Wayne's translation of *Faust Part 2*, Harmondsworth: Penguin Books, 1959, 269.

Sakharov's father—Dmitrii Ivanovich Sakharov—was an outstanding teacher of physics, who had educated several generations of teachers who, in turn, remembered him with gratitude and love. Dmitrii Ivanovich was the author of a physics textbook which was popular in its time, and a remarkable collection of physics problems, now undeservedly forgotten. After the death of Dmitrii Ivanovich several editions of this textbook came out under the editorship of his son.

The roots of the many qualities that together made Sakharov a true representative of the Russian intelligentsia—his democratism, modesty, respect for the opinions of others, his ability to remain faithful to his principles and convictions in all circumstances—may well lie deep in his family traditions.

Sakharov's path to science proved to be a difficult one. In the late autumn of the tragic year of 1941 he was evacuated to Ashkhabad together with other undergraduates and postgraduates of Moscow University.

After graduating from Ashkhabad University in spring 1942, Sakharov spent several months working at a tree-felling site in a remote rural region near the town of Melekess. In September of that year he was sent to a large military factory on the Volga, where he worked as an engineer. Shortly afterwards Sakharov's father received from his son the manuscript of an article on theoretical physics. The pages were covered with complex formulae and vector analysis symbols. Dmitrii Ivanovich showed this work to his friend A.M. Lopschitz, a mathematician and specialist in vector and tensor analysis. Lopschitz realised that the contents of the article were extraordinary and gave the manuscript to Tamm. Sakharov was subsequently summoned to Moscow, where from 1945 to 1947 he did postgraduate work at the Academy of Sciences Institute of Physics. From 1950 Sakharov worked at Arzamas-16. 'For the last twenty years I have worked incessantly in conditions of the utmost secrecy and the utmost tension, first in Moscow and then in a special Research Institute. At the time we were all convinced of the vital importance of this work for maintaining the balance of power throughout the world and were carried away by the sheer grandeur of it,' wrote Sakharov, speaking about this period of his life.

I shall always remember Zel'dovich's assessment of Sakharov's potential. During one of my last conversations with him he told me: 'I envy Andrei Sakharov. My brain is built to work like a well-maintained computer. But a computer only works if it is pre-programmed. Sakharov's brain writes its own programmes.'

In 1953, at the age of thirty-two, Sakharov was elected a full member of the Physics and Mathematics Division of the Academy of Sciences. Academician I.V. Kurchatov, who introduced him, told a meeting of the Division: 'This man has done more for the defence of our country than anyone else here.' When the vote was taken, Sakharov was elected on the first ballot. He was the youngest member of the Academy. For his contribution to the defence of the Soviet Union Sakharov won a State Prize, a Lenin Prize, and was three times nominated—in 1953, 1956 and 1962—a Hero of Socialist Labour.

Being a member of the top scientific echelons got Sakharov and other colleagues into unusual and even comic situations. Each of them was regarded as property under the protection of the state. The job of protecting them was done by armed bodyguards, who never left their side, and accompanied them to work, on walks, to the shops, and on visits to friends. It was entertaining to see Sakharov, plunged in thought, walking along in an armed convoy.

In 1950 Tamm and Sakharov conceived the idea of the magnetic thermo-isolation of plasma in order to produce a controlled thermonuclear reaction. According to Tamm, 'Sakharov not only devised the basic method which gave rise to hope for the implementation of thermonuclear reactions, he also did wide-ranging theoretical research into the properties of high-temperature plasma, its stability, and so on.' As a result of the efforts of a large group of Soviet scientists over many years, the 'Tokamak' system was created, which is now acknowledged to be the best. It is extremely close to the original idea of Sakharov and Tamm.

At about the same time Sakharov put forward the idea of magnetic cumulation. This new phenomenon involved the concentration of a magnetic field which allowed its original intensity to be increased by a factor of hundreds or even thousands. At its simplest, the magnetic-cumulative generator was a charge of explosive material placed outside a solenoid wound around a hollow metal cylinder. Before the charge was exploded, the solenoid created a primary magnetic field. When the charge exploded, the cylinder contracted. Because the magnetic current was preserved, the strength of the magnetic field increased by the same factor as the section of the cylinder reduced through explosive compression.

The first experiments in magnetic cumulation, carried out by R.Z. Liudaev, E.A. Feoktistova, G.A. Tsirkov, A.A. Chvileva, and others in spring 1952 confirmed the indisputable potential of this work. In the very first experiment the magnetic field increased twenty-five times over its original strength. Experiments conducted in the 1950s produced magnetic fields with an intensity of five million gauss. Later, using improved magnetic-cumulative generators, A.I. Pavlovskii's group created reproducible super-intensive magnetic fields in excess of 10 million gauss. At the present time, the development of various explosive-magnetic generators (explosion dynamos) has become a whole separate industry. Surge currents of up to three hundred million amps have been generated.

Theoretical physics was Sakharov's first love. He remained faithful to it all his life. 'The thing I love most of all in the whole world is residual radiation, which brings us information about the first moments of the Universe's existence.' Sakharov made a substantial contribution towards solving the fundamental problems of physics and cosmology. He did pioneering work on the fundamental problem of the construction of the universe—the question of why matter predominates over antimatter or, as the physicists put it, the problem of the baryon asymmetry of the universe. If the universe were symmetrical, the number of particles in it would be equal to the number of antiparticles; on collision they would be mutually annihilated, and instead of all the matter that surrounds us, and instead of you and me, nothing at all would exist except for some quanta of light.

How did this baryon asymmetry of the universe arise? Within the bounds of traditional conceptions there was no explanation. In a work written in 1967 Sakharov proposed the revolutionary idea of proton decay, the instability of the proton. The lifetime of this traditional 'building block' of the universe was posited here as greater by a factor of billions than the age of the universe. According to Sakharov's theory, this explains why a total mutual annihilation of matter and antimatter did not take place during the first instants of creation, and how instead there arose a small remnant of protons which proved sufficient for the formation of all the galaxies, stars, and planets.

This paradoxical and profoundly revolutionary conclusion, which bore on the very foundations of the microworld, was greeted quite sceptically by scientists. Several years later, however, the development of physics was proceeding in line with Sakharov's conception. Subsequent detailed calculations of the period of proton decay gave grounds for attempting to detect that decay experimentally and for the establishment of highly complex installations in various countries of the world.

Thus, one of the 'craziest' ideas in twentieth-century physics, put forward by Sakharov in 1967, had by the early eighties gone from the realm of pure theory into the realm of realizable experimentation.

Beginning in 1969, when he became a senior researcher in the Academy of Sciences Institute of Physics, Sakharov published some twenty works dealing with this question.

Fate led Sakharov to the creation of the hydrogen bomb. Unlike Fermi, he was unable to regard that project as merely 'interesting physics'. 'At first,' he recalls, 'I perceived the inevitable consequences of a hydrogen bomb explosion purely abstractly. My sense of special responsibility arose during test explosions, when you see scorched birds writhing on the burnt expanses of the steppe, when you see the shock wave blowing down buildings, when you smell the shattered brick and melted glass, and finally feel the explosion, when the shock wave sweeps across the steppe, flattening all the stalks in its path, and finally reaches you and knocks you over.'

This heightened sense of responsibility compelled Sakharov to analyze the consequences of using nuclear weapons and their effects on the incidence of oncological and hereditary diseases. In 1959 the publishing house Atomizdat put out a small anthology whose central article was Sakharov's 'Radioactive Carbon from Nuclear Explosions and their Non-Threshold Biological Effects.' Sakharov regarded uncontrollable mutations caused by nuclear testing as an additional factor in the deaths of hundreds of thousands of people. 'What is unique in the moral aspect of the problem is the utter impunity surrounding the crime insofar as that in each fatality it is impossible to demonstrate that the cause lies in radiation or in total helplessness of our descendants against the consequences of our actions.' In 1969 Sakharov transferred nearly all his savings, more than 130,000 roubles, to the Red Cross and a fund for building an oncological centre in Moscow. He was also one of the chief initiators of the 1963 Moscow Convention which banned nuclear tests in the atmosphere, in water, and in space.

It was a characteristic of Sakharov that he was utterly indifferent to the possible consequences to himself of his actions in defence of individuals or groups. During the late forties and early fifties an oppressive atmosphere developed which left its mark on the psychology and fate of many of the scientists at Arzamas-16.[28] In 1951 our personnel department discovered that the mathematician M.M. Agrest, who was down for promotion, had had a religious education and been certified as a rabbi. A decision to dismiss Agrest was quickly reached. In this situation the scientists at Arzamas-16 acquitted themselves in different ways. Some ceased greeting Agrest. Sakharov, on the other hand, gave him the use of his Moscow flat for many months.

After the Twentieth Party Congress the struggle to resurrect in the Soviet Union the science of genetics, which had been destroyed by Lysenko, became possible. This was largely done by physicists and mathematicians. On Kurchatov's initiative, a radiology department was set up in the Institute of Atomic Energy, while Academician M.A. Lavrent'ev established the Institute of Genetics at Akademgorodok, near Novosibirsk. At the same time Tamm organized a seminar on problems of molecular genetics and radiobiology at the Academy of Sciences Institute of Physics. Nevertheless, Lysenko's activities, which were ruinous for science and agriculture, not only were allowed to continue, but actually found support from the government. The decisive battle for genetics was not won until June 1964, with the elections of full members to the Academy of Sciences.

On my desk there is a yellowing sheet of paper almost a quarter of a century old. It is a transcript of the controversial session of the Academy of Sciences which took place on 26 June 1964. Sakharov's speech at that session illustrates his stance, and deserves more detailed treatment. Normally the confirmation of full academicians and corresponding members at a general meeting of the Academy of Sciences was a mere formality. Such meetings were attended by academicians from all fields of learning, and, as a rule, the decisions of the various departments were confirmed, since historians, linguists, chemists and biologists were fairly indifferent as to who, say, the physicists decided to elect to the places allotted to them. However, when the time came to confirm

---

[28] Another typically guarded reference, this time to Stalin's post-war repression in general and to his anti-semitic campaign (see note 6, p. 114) in particular.

the candidacy of N.I. Nuzhdin, who had been put forward by the biologists, things turned out differently. Nuzhdin was Lysenko's protégé and had received sixty-seven percent of the votes at departmental level. At the departmental meeting the first to speak was Academician V.A. Engel'gardt. In the best academic style, following P.L. Kapitsa, who had suggested that the number of references to an academic's work should determine the status of that academic, Engel'gardt asked Nuzhdin to withdraw his candidature, since he had not found a single reference to his work.

The second speaker was Sakharov. This proved to be the highlight of the meeting. 'As for me,' he said, 'I call on all academicians here present to cast their votes in such a way that the only votes in favour will be those of people who, together with Nuzhdin and Lysenko, bear responsibility for those difficult and shameful pages of Soviet science which are now, fortunately, coming to an end.' Tamm also spoke and, when the votes were counted, 114 were against Nuzhdin, with only 23 for him. This was the final blow to the anti-scientific ideas of Lysenko and his followers.

In the mid-1960s Sakharov became an active campaigner for human rights. In 1968 he addressed an urgent warning to all peoples of the world, to all progressive humanity, in the shape of his essay 'Reflections on Progress, Peaceful Co-existence, and Intellectual Freedom'. Here, for the first time, Sakharov identified and demonstrated the indissoluble link between the happiness and security of peoples, on the one hand, and the freedom and rights of the individual on the other.

Very soon after the appearance of this article, Sakharov was removed from secret work and dismissed from Arzamas-16. His colleagues in the Theoretical Department of FIAN demonstrated solidarity, compassion, and courage. On the instructions of Tamm, who was by then seriously ill, Feinberg went to see Sakharov and invited him to return to work at FIAN. Sakharov readily accepted his invitation. Opposition from the management and from Party sources was overcome. However, members of the Theoretical Department were continually harassed by Party organs. For example the local Party Secretary came into the department and shouted: 'How dare you maintain friendly relations with him! You should keep him at arm's length.' The district Party committee tried to 'persuade' by means of threats the secretary of the party group in the department, the eminent scientist E.S. Fradkin, a wounded and decorated war veteran, but

155

he stood firm. The head of the department, Academician V.L. Ginzburg was summoned before the praesidium of the Academy of Sciences and asked to sign the famous 'anti-Sakharov' letter. He refused.

'From 1970 onwards,' wrote Sakharov, 'the defence of human rights, the defence of people victimized by the political system, became my top priority.' What set Sakharov apart from many others was that for him there was no gap between conviction and action, between his word and the main strategy of his life.

In 1973, in an interview with a Swedish correspondent, Sakharov warned the world community of the danger and inadmissibility of unilateral disarmament by democratic countries. He thought it could provoke unforeseeable aggression. From that moment onward he became the target of a mass campaign of abuse, threats, and libel—the orchestrated 'wrath of the people'. 'Labour collectives', writers, and scientists united in attacking him. Sakharov's life, and the lives of his wife and children, were in danger. But Sakharov's public stance remained unchanged. When principles of public life were at stake, he devoted a huge amount of energy to the defence of prisoners of conscience, who were suffering for their political or religious convictions. He wrote letters, both to the Soviet authorities at various levels, and to international organizations; he attended trials, visited places of exile, announced hunger strikes, and convened conferences. In the course of this gruelling struggle there were few victories and many defeats, but he never relented. In the eyes of many people, in the East and in the West, Sakharov became a symbol of justice, a defender, and a last hope. He received a continuous stream of letters appealing for help. One of these bore the telling address: 'Moscow. Ministry of Human Rights and the Defence of Man. For the attention of A.D. Sakharov.' It was entirely fitting that he should have been awarded the Nobel Peace Prize in 1975.

Fifteen years before his time, Sakharov demanded full amnesty for political prisoners, freedom of the press, freedom to strike, freedom to choose one's place of residence within the USSR, freedom to leave the USSR, and to return, the independence and partial denationalization of state enterprises, and the introduction of a multiparty system.

In 1979 Sakharov spoke out against the deployment of Soviet troops in Afghanistan. The government's reaction was swift and unequivocal. He was stripped of all his government awards and exiled indefinitely, without trial, to the city of Gor'kii.[29] 'From the moment I was seized and brought to the prosecutor's office on 22 January 1980, I have been living in Gor'kii under arrest. There's a police guard on the doors of my apartment twenty-four hours a day, but you can't call it house arrest, because I'm not living in my own house, and you can't call it exile, because when you're exiled, you don't have guards at your door, and your contacts with visitors are not limited. Practically no-one, apart from my wife, is allowed to see me.'

Sakharov's colleagues in the Theoretical Department resolved to fight to allow him to remain a member throughout this time and to continue his scientific work, and to allow contact to be maintained with him. Ginzburg took these proposals to the praesidium of the Academy of Sciences and, when he had no success there, took the bold step of appealing to the Central Committee of the Party. At that time people were afraid even to pronounce Sakharov's name aloud. Ginzburg's proposals were sent on to higher authority, and there soon followed official consent on all points: Sakharov remained a member of the department, and scientific contacts with him were maintained. All told, during his years of exile forty-five 'man-visits' were made to him. Everyone considered these visits a great honour and a great pleasure. When there was a rumour that Sakharov was dying, Ginzburg rushed to see him. Fortunately, the rumour turned out to be false, and the meeting was a happy one for both men.

Sakharov was not exiled in perpetuity. On 15 December 1987 the militia presence was unexpectedly withdrawn from his apartment, and a telephone was installed in his room. The next day Mikhail Sergeevich Gorbachev telephoned him and invited him to return to Moscow.

On the first day after his return, despite a sleepless night on the train and the noisy welcome at the railway station, Sakharov went to the Theoretical department of FIAN. As luck would have it, it was a Tuesday, the traditional day for Tamm's seminar. The scheduled speaker began his paper with the words: 'As Sakharov has demonstrated.' At the seminar, which was chaired by E.L. Feinberg, there was an atmosphere of elation and joyful celebration. After the seminar Sakharov and his closest colleagues

---

[29] See Part 1, page 11, note 7.

gathered in the little study which had had his nameplate on the door throughout, and talked, and talked. No-one wanted to leave. That first day Sakharov spent about six hours in the Theoretical Department of FIAN. It was clear that this was his second home.

Sakharov was a winner of the Nobel Peace Prize, an honorary member of numerous foreign academies, and a People's Deputy. Scientists and statesmen, premiers and presidents, all visited him in his modest apartment. Receptions were organized in his honour, and the leaders of the superpowers listened to his opinions. In his simple and natural way, owing to his genius and the generosity of his heart, he became a spiritual leader of the contemporary world.

Scarcely had we put the final full stop on the manuscript of this book when the tragic news reached us: on 14 December 1989 Andrei Dmitrievich Sakharov died suddenly of a heart attack. Death is always unexpected, but the death of such a man is doubly so. 'Andrei Dmitievich was a great phenomenon, who transcended national boundaries. He went beyond what is preordained for every human being on earth. No-one now can strike his name from history, which he has entered as a great son of his people.' It would be difficult to add anything to the fine and prophetic words of Sakharov's friend Anatolii Marchenko.

LIEUTENANT-GENERAL PAVEL MIKHAILOVICH ZERNOV

KHARITON first encountered the name of Pavel Mikailovich Zernov, the organizer and first director of Arzamas-16, about a year before he actually met him. 'On 2 May 1945,' Khariton recalled, 'one week before the end of the war, a group of specialists, including some physicists, of whom I was one, landed at Tempelhof airfield in Berlin. From there the whole group travelled north-east from Berlin. Along the road at several places we saw signs with the words 'Zernov's factory'. When it became obvious in the course of our work that we would need to set up a special construction bureau and institute,[30] knowing full well that my own organizational abilities were unequal to the task, I asked Kurchatov for his advice. He recommended that I turn to the USSR Council of Ministers, which dealt with issues connected with our project, and request that a qualified director be appointed to deal with all the administrative work. A short time later I was summoned to a meeting in the Kremlin. There I was introduced to deputy minister of tank construction, General P.M. Zernov, whose factory was situated near Berlin. The meeting nominated Zernov to head our organization.'

Many years later one of Khariton's deputies remarked: 'It was a tremendous stroke of luck for our whole project that these two outstanding people met. Each of them was extremely valuable in his own right, but together they were pure gold, constituting that unity of scientific theory and engineering practice which ensures the success of large-scale scientific endeavour. From the very outset they worked together splendidly, and performed so harmoniously and amicably that we, their assistants, never thought of going to Zernov to corroborate instructions received from Khariton, or vice versa.'

What did Zernov do before he was appointed director of Arzamas-16? Where did he acquire such extensive experience as a major administrator and scientist? Here is the basic outline of his biography. He was born on 19 January 1905 in the village of Litvinovo, on the banks of the River Peksha in the Vladimir region. At the age of eight he went to a church school. The family was poor, but he did well at school. From 1918 to 1919 he worked as a hired farm hand. In 1920 he became an errand boy in a non-ferrous metals treatment plant in the neighbouring town of Kol'chugino. In 1926 he enrolled in the workers' faculty of the Plekhanov Institute of Economics. After graduating from there in 1929, he enrolled in the Bauman Higher Technical School in Moscow, where he specialized in internal combustion engines and from where he graduated with distinction in 1933. He was now twenty-eight years old. He remained in the postgraduate school, lecturing on thermodynamics, internal combustion engines, and other disciplines. Four years later, in 1937, he defended his doctoral dissertation, which was published and awarded a prize by the Moscow branch of the Komsomol. In 1938 he became chief

---

[30]These were both cover names for Arzamas-16 (see 'Introduction').

engineer and head of the Central Directorate of the Ministry, and the following year he became a deputy minister.

Here is one episode from his life at that time. At one of the plants under his jurisdiction an artillery tractor had been developed which, for some reason, was regarded as a 'wrecker's machine'. In those days the word 'wrecker' was popular, and highly dangerous.[31] Work on the tractor was halted. Zernov investigated the situation, organized factory testing of the tractor and, convinced that there was nothing wrong with the machine, took a bold and decisive step. He proposed to the Central Committee of the Party that two of the tractors should be tested by driving them from Stalingrad to Moscow. His extraordinary boldness and decisiveness were fully justified. Under observation by military representatives, both vehicles reached Moscow safely and were accepted as matériel for the Soviet army.

At the age of thirty-five Zernov was appointed chairman of the All-Union Committee of Standards. Zernov did not like office work, but even here his distinctive approach could be seen. He described this brief period of his life thus: 'The War Department proposed increasing the standard thickness of soldiers' box-calf boot soles from four to five millimetres. That, of course, required additional use of leather. After making the necessary calculations and the corresponding cutting patterns for the boots, I proposed compensating for the increased use of leather by lowering the boot tops by three centimetres. The supplies department of the Ministry of Defence did not agree to this. The issue went to the Council of Ministers, at one of the meetings of which S.M. Budennyi criticised the proposal from the Committee of Standards. I insisted on my point of view. At that point six servicemen were invited into the room, three of them wearing boots cut to the new pattern. Stalin turned to Budennyi and said: "Show us which boots you like best." "I can't tell the difference," replied Budennyi. "In that case the proposal of the Committee of Standards is accepted." That was the end of the matter.'

As deputy minister of tank construction, Zernov was in charge of tank factories and did a variety of tasks for the State Defence Committee. At the very beginning of the Second World

---

[31]'Wrecking' and 'wrecker' were watchwords in the prosecution of alleged 'counter-revolutionary sabotage' in agriculture and industry, often figuring in the show trials of the 1930s.

War, as the authorized representative of the State Defence Committee, he increased tenfold the production of heavy machine-guns at a factory near Moscow. He accelerated tank production and organized the evacuation of tank factories to the East and was soon in charge of assembly-line tank production in Cheliabinsk. He himself left Khar'kov on the last plane out. There were no seats in the cabin, so he travelled in the bomb-bay. In June 1942 Zernov was put in charge of tank production in Stalingrad. The tanks went straight from the factory gates to the front. In December 1943, still in Stalingrad, he headed a commission charged with restoring the city and its industries.[32]

Zernov had gone from farm hand to deputy minister before becoming director of Arzamas-16, one of the largest research centres in the country. Here his knowledge and ability were as much in evidence as before. He was called upon to solve complex and very diverse problems. Creating a strong base for experimental production, organizing a highly qualified production staff, mastering new and complex technological processes, building houses for newcomers, ensuring food supplies when food was rationed throughout the country, organizing farms, a medical centre, a theatre and a cinema—this is nowhere near the complete list of the tasks for which he was responsible. They called not just for a good organizer, but for an extraordinary human being with special personal qualities.

Restrained and attentive, very organized and decisive—that is how he is remembered by everyone who knew him. He taught us a great deal. I have never forgotten one discussion in particular, on the subject of false ambition. I had come to his office for some reason or another. Our conversation quickly turned to the disputes which were dogging the first years of our existence. 'Do you know what false ambition is?' he asked me. 'False ambition is when someone argues without real cause, argues simply to boost his own prestige and ambitions. You have to know not only how to argue your point, but also how to acknowledge that you're in the wrong.'

Zernov was both very exacting, and very fair. He valued and respected his workers. Housing was a problem, but Zernov always found ways to provide housing for his workers and their families. He could walk into a kitchen or a refectory and show

---

[32] The Battle of Stalingrad (July 1942–February 1943) left the city in ruins.

everyone how best to make tea. He took the trouble to make sure that the bread we ate was good and tasty. Once I saw on his desk, of all things, a length of sausage. 'Want to try some?' he asked. 'We've decided to organize our own sausage production. We're going to have sausage of every description.'

Once, a couple of workers came to have lunch in the refectory at four o'clock, that is, after the refectory was officially closed. They were refused service. Zernov, who was in the refectory at the time, came over, saw what the matter was, and gave instructions that they be given lunch. At the same time, he made arrangements for cooks to be on duty around the clock. There is nothing surprising in this, since at this time we seldom worked less than twelve to sixteen hours per day. People were often delayed on the test-sites until dawn.

Zernov also instituted a system whereby any request would be considered on the day it was made. When he discovered that bureaucratic procedures were hampering the delivery of materials, Zernov convened a special meeting in which not only departmental heads, but also representatives from the accounts and supplies departments, took part. 'Tell me, has the incidence of theft decreased since requests have had to be completed in triplicate, with six signatures for every component?' he asked the chief accountant. 'Well, no, Pavel Mikhailovich. That's just the system.' 'I'm sick and tired of these systems. Let's try to make do with one request and two signatures: one from the person requesting materials or equipment and the other from the storekeeper.' That is what they did. A couple of months later he again convened a meeting and put the same question and, as might have been expected, got the same answer.

Zernov was sometimes criticized because we were not working 'correctly', that is to say, without paperwork, without official instructions, without an epidemiological station and other similar services. The absence of paperwork was fine—there was all the more time for real work. When we were inspected, Zernov would say: 'We took on the best specialists in the country to work here, who know all there is to know, who have to provide for every eventuality in their work and who are responsible for everything they do. Do inspectors know more than they do?'

In 1948 I happened to go into Zernov's office while he was talking to some builders. On his desk there was a general plan for the future town of Arzamas-16. 'Here there'll be a wide avenue.

We'll probably call it Oktiabr'skii Prospect.'[33] Zernov traced a line from south to north through an expanse of forest. 'We'll have buses running up and down it, and eventually trolleybuses.' All around was impenetrable forest, and he was planning the building of a town. It seemed a distant dream at the time.

Zernov lived a very intense life. He worked for eleven years without taking a holiday. In his autobiography he relates an incident which took place before the war. While at a meeting in the Kremlin, he suddenly lost consciousness. He did not come to until twenty-nine hours later. It turned out that he had had no sleep for seven days.

In 1950 Zernov suffered a serious heart attack which put him out of action for seven months. In 1951 he moved to Moscow. Khariton recalled: 'Zernov worked for us for four or five years. Yet we all had the feeling that we had worked with him for at least fifteen years, so much did we experience psychologically in that brief period which was crammed with events both large and small.'

Zernov died on 7 February 1964 while deputy minister of medium machine-building.[34]

Major-General Boris Glebovich Muzrukov

Boris Glebovich Muzrukov...As I say this name, my mind pictures a very tall, well-built man. The keen look of his transparent, nearly grey-blue eyes beneath his beetling brows created an impression of severity. In reality, he was a kind, responsive, and compassionate man.

A graduate of the Leningrad Polytechnic Institute, chief metallurgist at the Kirov factory in Leningrad, one of the country's leading arms experts, who had organized production of the legendary T-34 tank and the IS self-propelled gun, one of the first people to be twice awarded the title of Hero of Socialist Labour, Muzrukov belonged to a glorious pleiad of outstanding

---

[33]This is, in fact, the name of one of the main streets of Arzamas-16/Sarov.

[34]The Ministry of Medium Machine Building was, from 1953, the cover name for the Ministry responsible for all secret nuclear installations in the Soviet Union. Now renamed the Ministry of Atomic Energy. (See 'Introduction'.)

leaders of industry, whose talent and selfless labours, both in the pre-war years and, especially during the war, ensured our victory.

Muzrukov's style of work did not square with the usual conception of directors of his standing. He was not afraid to take responsibility for decisions, but differed from the authoritarian type of director in his manner of dealing with subordinates. He could not stand rows when dealing with conflict situations, and always spoke in soft measured tones. He laughed rarely, almost silently, with a funny childlike way of screwing up his eyes, while his smile was somehow apologetic, especially at meetings, as if he were embarrassed at having allowed himself a little laugh during a serious discussion. He was demanding, and kept abreast of things at all levels.

In 1956 I was on business in Sverdlovsk. Although it was some years since Muzrukov had been director of the Uralmash factory, his name was on everybody's lips there. People called him 'Tsar Boris', no less.

Muzrukov himself described this important period of his life thus:

> During the pre-war years I, together with other young specialists, was sent abroad by the government to study the latest technology. I was assigned to the Russian naval commission in the town of Livorno, Italy. The firm of Ansaldo had a contract with the Soviet Union to build a high-speed destroyer. It seems strange that Fascism was arming the Soviet navy, but Mussolini took the view that the two countries had no ties, and no common borders, and that therefore the Italians could show the Russians whatever they had, and build them whatever they wanted. Our government took advantage of Mussolini's view. I was charged with overseeing quality control of everything going to the USSR, particularly the ship's engine assembly. Eventually this ship was named the *Tashkent.*

In 1947, in response to developments in the atomic weapons project, Muzrukov's destiny took a sharp turn, and the whole of the rest of his life was linked with this new branch of industry. He was appointed director of a major new industrial combine concerned with mining and refining raw materials for the nuclear industry. He stayed here until 1949, when he was transferred to work in a ministry.

In 1955 Khariton met Muzrukov in Moscow. In the course of their conversation, Khariton gathered that ministry work was not to Muzrukov's taste, and that he was keen for work with people, which he found more exciting and to which he was accustomed. He eagerly accepted an offer to become director of Arzamas-16.

The character and spirit of a person are best revealed not when things are flowing smoothly, but in those relatively rare situations when quick and accurate decisions need to be made. In two decades of work under Muzrukov, there were a number of such instances. I would like to relate one of them in detail.

It was late January 1960. I'd no sooner arrived for work than the phone rang. It was Muzrukov's secretary. 'Boris Glebovich asks you to drop everything and come and see him immediately.' I did not like orders of this sort, but you have to do what the director says. I finished off what I was doing, and fifteen minutes later I was at the factory. I could see straightaway that something unusual had happened. There were additional guards on the entrance. Despite a sharp frost, all the workers and technical staff were standing outside. The first to address me was the head of the political department Aleksandr Stepanovich Silkin. 'It's most urgent that you determine whether we'll have to evacuate the residents of the town. Any evacuation takes time.' Soon Muzrukov came out of the building. 'Ah, professor, let's go and get changed. You're just the man we need.' I got changed, and went up to the dosimeter staff's room. They were periodically switching on the dosimeters, which would give out a melodic tone, signifying increased radiation. The level in the work area was roughly three times higher than allowable, but there was no need to evacuate the town.

The main feature of this room was a line of organic glass boxes. The boxes were evidently not very hermetic, and radioactive gases and sprays were seeping into the room through the gaps. I remembered an old remedy for situations like this—plasticine. I turned to Muzrukov. 'Boris Glebovich, please give the order to get hold of about twenty kilos of plasticine right away.' Three hours later we had the plasticine. Using radiation counters we began locating the gaps and sealing them. The radiation count decreased noticeably. That evening I had another idea: in the dark it should be easy to make out argon luminescence (argon was the basic gas in the boxes). Sure enough, as soon as the lights were turned off, the boxes began glowing light-blue.

Where there were gas leaks, the glow was more intense. It was a remarkable spectacle.

When gas samples had to be taken, glass samplers were first pumped out and then attached to the line of boxes with the help of taps. After the sample had been taken, the samplers were disconnected. Thick vacuum rubber was used to connect the samplers. The first attempt, by Davidenko, to take a sample was a failure. Vera Sof'ina, who was among those present, turned to Muzrukov: 'Boris Glebovich, ask Veniamin Tsukerman to do it.' 'But he can't see anything.' 'That doesn't matter. When I have to get a thick vacuum rubber onto a glass connection, he's the only one I trust.' Muzrukov turned to me: 'Professor, is it true you can do that?' 'There was a time when I could. I think I could still manage it.'

They got me about five samplers. I rejected certain ones after gauging their weight by feel. Then I moistened the connector with alcohol, and successfully fitted the rubber onto it.

'You see, Boris Glebovich, like most sighted people, you underestimate the ability of the blind to handle delicate apparatus.'

Rather touchingly, Muzrukov would help me to put on my protective clothing. I usually forgot the boots, at which point Muzrukov, who had been watching the whole procedure, would bend down, and say: 'Professor, you've forgotten to put on your boots. Let me give you a hand.'

At the same time, the laboratory was hard at work researching the absorption and secretion of gases by various metals. For seven days and nights our people struggled stubbornly to reduce the level of radioactivity in the air of the main building. By the end of this period it was only twenty-five percent above the admissible level. Throughout this entire period Muzrukov left the site only for two or three hours at a time at night. No matter what the radiation level, he was always in the most critical areas. We used to tell him: 'Radioactive gases and sprays don't know that you're the director. They have exactly the same effect on the machine attendant and the director.' These remonstrations did not help much. In the end, tests showed that Muzrukov had taken a dose of 0·1 conventional curies, compared with the 0·025 taken by the author of these lines. A 'man of the old school', Muzrukov thought that the director of the subdivision, who bore full responsibility for safety procedures, ought to be present personally during work in the most dangerous areas. Apart from the radioactive gases, the cold caused us a great deal of trouble. In order to purify the air in the working areas as quickly as possible, we used to open the entrance gates to the building several times a day. The outside air temperature ranged from -25°C to -30°C. In the end we collected about thirty percent of the radioactive gas. Afterwards Muzrukov caught a cold and developed pneumonia. The tuberculosis from which he had suffered in his youth had its effect, and he was ill for about two months.

In 1967, for the fiftieth anniversary of the October Revolution a 'Club of Interesting Meetings' was formed in one of the research subdivisions at Arzamas-16. It quickly became popular, and attracted the town's intellectuals. Soon the club had outgrown the subdivision and become both a centre for the promotion of scientific knowledge, and a centre for leisure. It did not, however, have any permanent premises, and met wherever it could in the evening. We felt like wandering musicians, who lacked only a horse and a phaeton for moving our effects around. We turned to Muzrukov for help. He appreciated the initiative we had shown and, despite doubts and objections from certain persons in high office, organized the construction of a purpose-built building with a 270-seater auditorium, a spacious lobby, a sound system, and a film projector. At first it was named the Café Science, but was soon renamed the 'House of Scientific and Technical Workers'. Its first director, Nina Ivanovna Kuz'mina remembers: 'Boris Glebovich taught me not to strive to express myself, but for others to express themselves. The task of the director was to create the conditions for this. In a club like this the main thing is always the spirit of creativity. If I had any questions, I was to go and see Muzrukov between 7.40 and 7.45 and we would sort them out together. This helpful attitude enabled the building to be completed to a tight schedule. It was opened on 3 February 1973. Unquestionably, without Muzrukov's active support, the building would never have happened. As it is, it continues to function actively to this very day.'

Muzrukov was a very modest man. L. Riabev recalls a letter from Muzrukov written in response to plans for celebrating his birthday:

> *To the Secretary of the Party Organisation of the Institute, the Director of the Institute, the Chairman of the Commission:*
> It has come to my notice that, in view of my seventieth birthday, an anniversary celebrations commission has been

formed. As a communist, I object in principle to any anniversary festivities, and I therefore request that no measures are taken on my behalf whatsoever.

17 September 1974 *B.G. Muzrukov.*

The birthday celebrations were cancelled. However, Muzrukov's friends did prepare a sort of present for him on the day—a commemorative evening in the 'Scientists' House', entitled 'The Urals Forge Victory'.

Muzrukov's kindness to all living things accelerated his end. On the wall of the house where he had lived for fifteen years was a bird-table. Everyday he used to put grain or sunflower seeds on it. On 15 February 1978, a day of black ice, he went to the bird-table, slipped and fell. He never recovered.

EPILOGUE

TIME is the supreme judge, at once both prosecutor and advocate. Among those who will read these lines, there will be some who criticize them for their form or content. Our contemporaries will rebuke us for failing to mention many remarkable directors, researchers, and craftsmen, for over-emphasizing the role of some scientists, while ignoring others.

Can we offer anything in our defence against these largely justified reprimands? We have tried, to the best of our abilities, to follow the principle of 'not one word which goes against your conscience'. However, personal sympathies and affections, the limited field of vision of a participant in these events—these are obstacles in any historical narrative.

While travelling along the Black Sea coast of the Caucasus, on Mount Iverskaia, near Novyi Afon, we found a remarkable inscription among the ruins of an old Russian monastery. 'We were just the same as you. You will be just the same as us.' The anonymous monk invested these few words with a powerful meaning. In those words is the eternal cycle of life, which is subject only to a single argument, an argument not contingent on men—the argument of Time.

Today is our last day of work on this book. We are sorry to have to part with it. But every endeavour has to come to an end. And there go our two co-authors again—two black-and-white, red-crowned woodpeckers. Our forest radio-operators are back at their post, and have resumed tapping out their signals.

AFTERWORD

by Iulii Borisovich Khariton

I MET the authors of this book in August 1942 when V.A. Tsukerman came to see me to tell me about the method which he had devised for the high-speed X-ray photography of explosion and detonation processes. I liked both the daring proposal and its author, a man fascinated by science.

Veniamin Aronovich was the first major experimenter I brought in to work with me when I was given the job of creating what was then called a branch of Laboratory No. 2 of the USSR Academy of Sciences.

The scientific, inventive, and public activities of Veniamin Aronovich all demonstrated the broad range of his interests, his tenacity, the boldness and imaginativeness of his conceptions, and his acute sense of feasibility, and also his creatively inspired and intense, ceaseless labour. About a hundred publications, ten monographs, sixty inventions, tens of millions of roubles in revenue, fifty trained Doctors of Science—that is the sum of his work in science and technology.

One constantly marvelled at the kindness of Veniamin Aronovich, his boundless love of people, his active compassion. In the final analysis these are the highest criteria in evaluating a personality.

He holds the titles of Hero of Socialist Labour and Honoured Inventor, is a laureate of the Lenin Prize and four State Prizes, and has many other medals and decorations.

It is hard to imagine how all this fantastic volume of work could have been done by a man who cannot see. Despite this grave disability, Veniamin Aronovich has accomplished so much that it seems appropriate to call his life a heroic feat.

GLOSSARY OF RUSSIAN TERMS

Akademsnab a portmanteau word comprising the Russian words for 'Academic' and 'supply'.
Balaganchik literally: 'Fairground booth'.
bogatyr a heroic figure from Russian folklore.
burzhuika a primitive type of heating stove (literally 'bourgeois woman').
dacha type of out-of-town house owned by many Russians.
FIAN Russian initials for 'Institute of Physics of the Academy of Sciences.
fortochka a type of fanlight.
FSB Russian initials for Federal Security Service. Post-1991 name for Russian security police.
Gazik a car made at the GAZ factory. The initials stand for 'Gor'kii Automobile Factory'.
glasnost policy of transparency in government and, ultimately, of free speech, initiated by M.S. Gorbachev.
kasha a type of semolina.
KGB Russian initials for Committee of State Security. Name given to Soviet security service 1953-91.
Komsomol portmanteau word comprising the Russian words for 'Communist Union of Youth'. Usually translated as 'Young Communist League'.
likbez a portmanteau word comprising the Russian words 'liquidation' and 'illiteracy'.
Moskvich a type of car. Literally 'Muscovite'.
NKVD Russian initials for People's Commissariat for Internal Affairs. Name given to Soviet security service 1934-43.
nomenklatura in the USSR, members of the 'charmed circle' of people eligible for appointment by the Communist Party to lucrative posts.
perestroika policy of reform initiated by M.S. Gorbachev. Literally: 'restructuring'.
TASS Russian initials for Telegraph Agency of the Soviet Union, the official (and only) Soviet news agency.
Pobeda a type of car. Literally 'Victory'.
Uralmash a name derived from the Russian word for 'Urals' and 'machine'.
VNIIEF Russian initials for 'All-Russian Research Institute of Experimental Physics', the post-Soviet name for Arzamas-16.

BIBLIOGRAPHY

Books

In English or German

Davies, R.W., *Soviet History in the Yeltsin Era*, London, 1997.
Heinemann-Gruder, A. *Die sowjetische Atombombe*, Westfälisches Dampfboot, 1992.
Holloway, D., *Stalin and the Bomb*, Yale University Press, 1994.
Jungk, R., *Brighter than a Thousand Suns*, New York, 1958.
Marples, D.R., Young, M.J., *Nuclear Energy and Security in the Former Soviet Union*, London, 1997.
Pursglove, M. (translator), *Life or Hearing*, Reading: Bulmershe Resource Centre, 1990. (translation of the chapter on Irina Tsukerman from the book by Krainin, V., and Krainina, Z., below).
Rhodes, R., *Dark Sun*, New York and London, 1995.
Rhodes, R., *The Making of the Atomic Bomb*, New York, 1986.
Sakharov, A., *Memoirs*, London, 1990.

In Russian

Beriia, S., *Moi otets—Lavrentii Beriia*, Moscow, 1991.
Feinberg, E.L. et al. (eds.), *Vospominaniia o I.E. Tamme*, Moscow: Nauka, 1986.
Gubarev, V.S., *Arzamas-16*, Moscow, 1992.
Khariton, Iu.B., Smirnov, Iu.N., *Mify i real'nost' sovetskogo atomnogo proekta*, Arzamas-16, 1992.
Kochariants, S.G., Gorin, N.N., *Stranitsy istorii iadernogo tsentra Arzamas-16*, Arzamas-16, 1992.
Krainin, V., and Krainina, Z. (i.e. Tsukerman V.A. and Azarkh, Z.M.), *Chelovek ne slyshit*, Moscow: Znanie, 1984. 2 ed. 1987.
Makeev, N.G. (editor), *Fizika i tekhnika*, Sarov, 1996.

Bibliography

Articles

Al'tshuler, L.V., 'Tak my delali bombu', *Literaturnaia gazeta*, 23 (5297), 1990, 13.

Al'tshuler, L.V., 'Vo vsem mne khochetsia doiti do samoi suti', *Gorodskoi kur'er* (Arzamas-16), 5 (215), 21 January 1993, 6.

Gubarev, V.S., 'Taina sekretnogo cheloveka', *Ogonek*, 33, 1993, 18-23.

Khariton, Iu., and Smirnov, Iu., 'Otkuda vzialos'' i bylo li nam neobkhodimo iadernoe oruzhie', *Izvestiia*, 21 July 1994, 7.

Kolesnikov, A., 'Akademik Gol'danskii "Vzryv atomnoi bomby spas sovetskuiu fiziku"', *Ogonek*, 33, 1993, 23-24.

Pestov, S.,'Tainy atomnoi bomby', *Argumenty i fakty*, 40, 1989, 5-6

Stiazhkina, T., 'Roskosh'' chelovecheskogo obshcheniia', *Gorodskoi kur'er* (Arzamas-16), 26, 1993, 3.

Svetitskii, K., 'Zakrytyi gorod khochet zhit'' spokoino', *Izvestiia*, 19 September 1995, 5.

Volkova, G., 'Uznik XX s"ezda', *Obshchaia gazeta*, 34 (162), 1996, 9.

Franchetti, M., 'Happy Families in Stalin's Hellhole', *The Sunday Times*, 19 April 1998, 21.

Meek, J.,'Poverty puts nuclear materials on the market', *The Guardian*, 15 August 94.

Pursglove, M., 'The Genre of Silence: Iurii Nagibin's Zamolchavshaia vesna', in Hunter, W.J. (ed.), *The Short Story: Structure and Statement*, Exeter: Elm Bank, 1996, 159-72.

Rich,V., 'Secret cities get dollars', *Times Higher Education Supplement*, 1354, 16 October 1998, 10.

Tucker, A., (obituary) 'Yuli Borisovich Khariton', *The Guardian*, 20 December 1996, 13.

— (obituary) 'Yuli Khariton', *The Daily Telegraph*, 20 December 1996, 25.

— (obituary) 'Yuli Khariton', *The Times*, 20 December 1996, 19.

Translation

Nagibin, Iu., 'The Spring that Fell Silent', in *Lateral Moves*, 20, 1996, 30-37 (translated by M. Pursglove).

WHO'S WHO and INDEX OF PERSONAL NAMES

ADAMSKII, Viktor Borisovich (b. c.1920), scientist xix
AGREST, Mattes Mendelevich (c.1920-89), mathematician 147n., 154
AITMATOV, Chingiz Torekulovich (b. 1928), prose writer 143
AKHMATOVA, Anna Andreevna (1889-1966), poet xix
AKIMOV, Nikolai Pavlovich (1901-68), theatre director 139
ALEKSANDROV, Anatolii Petrovich (b. 1903), Academician 116
ALEKSANDROVICH, Konstantin Vital'evich, son of V.A. Aleksandrovich 81
ALEKSANDROVICH, Vitalii Aleksandrovich (1904-59), chemist 53, 66, 68, 79-82, 92, 97
ALEKSEEVA, Rufina Nikolaevna, librarian 64
AL'TSHULER, Boris L'vovich (b. 1939), physicist and civil rights campaigner. Son of L.V. Al'tshuler 24
AL'TSHULER, Lev Vladimirovich (b. 1913), physicist xiv, xvii, xix, xxxi-xxxii, 5, 8, 9, 10. 15, 17, 21, 24, 29, 37, 46, 47, 50, 51, 53-5, 62, 68, 72, 74, 78, 83-5
APIN, Al'fred Ianovich, radiochemist 53, 66
ARTSIMOVICH, Lev Andreevich (1909-73), nuclear physicist, Academician 133
ASTON, Francis William (1877-1945), English chemist. Nobel Prize for Chemistry 1922 127
AVDEENKO, Aleksandr Ivanovich, radiophysicist 10-2, 20-1, 23, 24-5, 27-8, 39-40
AVDEENKO, Liudmila Stepanovna, wife of A.I. Avdeenko 29
AVENARIUS, Richard, German philosopher (1843-96) 147n.
AVERBAKH, Mikhail Iosifovich (1872-1944), eye specialist 17-18
AZARKH, Moisha Girshevich (Matvei Grigior'evich), father of Z.M. Azarkh xx
AZARKH, Rakhil Zalmanovna (née Rubashova) (d. 1958), mother of Z.M. Azarkh xx
AZARKH, Zinaida Matveevna (Zina) (b. 1917), wife of V.A. Tsukerman x, 12-17 and *passim*
BAKANOVA, Anna (Ania) Andreevna (1921-79), physicist 53, 61
BAKH, Aleksei Nikolaevich (1857-1946) biochemist, Academician 28
BAKH, Natal'ia Alekseevna, chemist. Daughter of A.N. Bakh 28
BAKHMETEV, Evgenii Fedorovich (c.1900-44), aeronautical engineer 6-9, 18
BALDIN, Nikolai Andreevich, doctor. Chief medical officer of municipal hosptal at Arzamas-16 70
BARSKAIA, Elena Mikhailovna, librarian xxii, 64-5
BEDNOVA, Galina Napoleonovna (1926-88), engineer at Arzamas-16 87
BELIAEV, Aleksandr Fedorovich, explosives expert 35, 59, 129
BELKIN, Nikolai Vasil'evich (1937-98), physicist at Arzamas-16 56

Index of names

BERIA, Lavrentii Pavlovich (1899-1953). Secret Police Chief 1938-53. In charge of nuclear programme. Executed after death of Stalin xii, xvii-xviii, xxvi, 55, 68, 84
BESSARABENKO, Aleksandra Aleksandrovna, wife of A.K. Bessarabenko 86
BESSARABENKO, Aleksei Konstantinovich (d. 1960), deputy director of Arzamas-16 85-7, 108
BESSARABENKO, Nikolai Konstantinovich, brother of A.K. Bessarabenko 86
BIRENS, Mira Iakovlevna (d. c.1950), mother of Iu.B. Khariton. Real name: Burovskaia 122
BLÉRIOT, Louis (1873-1936), French aviator 56
BOCHVAR, Andrei Anatol'evich (b. 1902), metallurgist 115, 139-40
BODENSTEIN, Max Ernst August (1871-1942), German chemist 126
BONCH-BRUEVICH, Mikhail Aleksandrovich (1888-1940), radiophysicist 11
BONDAREV, head of Evening Institute 9
BORISOV, Sergei Ivanovich, linguist 65
BORODIN, Aleksandr Porfir'evich (1833-87), composer 112n.
BRAZHNIK, Militsa, physicist 53
BRISH, Arkadii Adamovich (b. 1917), physicist 53, 57-8, 73-4, 87-90, 96
BRISH, Liubov Moiseevna, wife of A.A. Brish. Assistant to M.V. Dmitriev 96
BUBNOVA, Varvara Dmitrievna (b. 1886), painter xxi
BUDENNYI, Semen Mikhailovich (1883-1973), Marshal of the Soviet Union 160
BUDKER, Gersh Itskovich (1918-77), nuclear physicist 114
CHADWICK, James (1891-1974), English physicist. Discoverer of the neutron. Nobel Prize for Physics 1935 127-9
CHAGALL, Marc (Mark Zakharovich) (1887-1985), painter 1, 5
CHAVCHAVADZE, Nina Aleksandrovna (1812-57), wife of A.S. Griboedov 31
CHERNYSHOV, Vladimir Konstantinovich, physicist, mathematician 101
CHETVERIKOV, Sergei Sergeevich (1880-1959), biologist 9
CHOPIN, Frédéric (1810-49), Polish composer 139
CHUDAKOV, Evgenii Aleksandrovich (b. 1921), engineer, Academician 11
COCKCROFT, John Douglas (1897-1967), British scientist 112n.
CRICK, Francis Harry Compton (b. 1916), British scientist. Nobel Prize for Medicine 1962 (with James Watson and Maurice Wilkins) 146
CURIE, Irène (1897-1856), French chemist. Nobel Prize for Chemistry 1935 (with her husband Frédéric Joliot-Curie) 128, 131
DAL, Vladimir Ivanovich (1801-72), lexicographer, ethnographer, and prose writer 137
DANKOV, physicist at Kazan University 25
DAVIDENKO, Viktor Aleksandrovich (d. 1983), physicist xxxi, 66, 90-5, 104, 166
DETNEV, Vasilii Ivanovich, security official 68

DIESEL, Rudolf (1858-1913), German engineer 28
DIRAC, Paul (1902-84), British physicist. Nobel Prize 1933 127
DMITRIEV, Mikhail Vasil'evich (1918-62), chemist xxxi, 95-8
DMITRIEV, Nikolai Aleksandrovich (born c.1925), mathematician xvi, 54
DOBROTIN, Nikolai, physicist 26
DUDINTSEV, Vladimir Dmitrievich (1918-98), prose writer xx, 85
DUKHOV, Nikolai Leonidovich (1904-64), mechanical engineer 78
EDISON, Thomas Alva (1847-1931), American scientist 8
EINSTEIN, Albert (1879-1955), Swiss-American theoretical physicist. Nobel Prize for physics 1921 46, 146
ENGEL'GARDT, Vladimir Aleksandrovich (1894-1984), biochemist, Academician 155
ETINGOV, E.A., radio engineer 57
EULER, Leonhard (1707-83), Swiss mathematician. Director of St Petersburg Academy from 1766. Blind from c.1766 18
FEDOROV, Mikhail Ionovich, barber at Arzamas-16 58-9
FEINBERG, Evgenii L'vovich (b. 1912), theoretical physicist. Student of I.E. Tamm 147, 155, 157
FEOKTISTOVA, Ekaterina Alekseevna (born c.1920), physicist 152
FERMI, Enrico (1901-54), Italian-American physicist. Nobel Prize for Physics 1938 149, 153
FILATOV, Vladimir Petrovich (1875-1956), eye specialist 18, 106
FLEROV, Georgii Nikolaevich (1913-90), nuclear physicist xxv, xxix, 64, 78, 105, 137
FRADKIN, Efim S., (b. 1924), theoretical physicist 155
FRADKIN, Mark Grigor'evich (1914-90), pianist and songwriter 2
FRANCHETTI, Mark, British journalist xii
FRANK, Il'ia Mikhailovich (b. 1908), experimental physicist, Academician 26, 29
FRANK-KAMENETSKII, David Al'bertovich (c.1915-c.1972), physicist xxxi, 54, 64, 74, 103-5
FRENKEL, Iakov Il'ich (1894-1952), physicist 125
FRISCH, Otto Robert (1904-79), Austrian physicist. Nephew of Lise Meitner xxix, 131-2
FUCHS, Klaus (1911-89). British nuclear scientist of German extraction, who spied for the Soviet Union xxx
GALYNKER, Izrail Solomonovich (1909-67), colleague of Veniamin Tsukerman xx, 3n.
GANDEL'MAN, G.M., theoretical physicist 54
GARCIA LORCA, Frederico (1899-1936), Spanish poet xxi
GAVRILOV, Viktor Iulianovich (1918-73), biophysicist 54
GEVELING, professor at Zhukovskii Air Force Academy 6
GINZBURG, Vitalii Lazarevich (b. 1916), physicist, Academician 9, 156-7
GODUNOV, Sergei Konstantinovich (b. 1929), mathematician 83
GOETHE, Johann Wolfgang von (1749-1832), German writer 148
GOLUBEVA, Liudmila, civilian at Arzamas-16 xxi, 69
GOL'DANSKII, Vitalii Iosifovich (b. 1923), chemist, Academician xiv, xvii, xix, xxi, xxiii-xxvii

Index of names

GORBACHEV, Mikhail Sergeevich (b. 1931), Soviet politician. Head of State 1989-91. Architect of *perestroika* and *glasnost* xiv, xvi-xvii, 27n., 157
GORKY, Maxim (1868-1936), dramatist and prose writer 139
GRANIN, Daniil Aleksandrovich (b. 1919), writer xx
GRIBOEDOV, Aleksandr Sergeevich (1795-1829), poet, dramatist, and diplomat 31
GRIEG, Edvard Harderup (1843-1907), Norwegian composer 45, 139
GUBAREV, Vladimir Stepanovich, Soviet journalist xiii-xiv
GUMILEV, Nikolai (1886-1921), poet xiv
HAHN, Otto (1879-1968), German chemist. Nobel Prize for Chemistry 1968 xxix, 132-3
HAWKING, Stephen William (b. 1942), British physicist 143n.
HAYDN, Franz Joseph (1732-1809), Austrian composer 139
HEISENBERG, Werner Karl (1901-76), German physicist and philosopher. Nobel Prize for Physics 1932 133
HEMINGWAY, Ernest (1898-1961), American writer xxi
HOLLOWAY, David (b. 1943), Irish-born author, xvii
IGNAT'EV, Ivan Ivanovich, glassblower 110
IOFAN, Boris Mikhailovich (1891-1976), architect 13
IOFFE, Abram Fedorovich (1880-1960), physicist xii, 37, 43, 101, 123, 126, 130
IUR'EVA, Nina Danilovna (b. 1927), assistant to Davidenko xxii, 90-2
JOLIOT-CURIE, Frédéric (1900-1958). Husband of Irène Curie 128, 131
KALININ, Mikhail Ivanovich (1875-1946), Soviet politician. Titular Head of State 1922-46 44
KANUNOV, Mikhail Alekseevich, technician 107-10
KAPITSA, Petr Leonidovich (1894-1984), physicist. Nobel Prize for Physics 1978 xii, xvii, 41, 43, 126-7, 130, 155
KARADZHA, F.N., radiography expert 8
KELDYSH, Liudmila Vsevolodovna, mathematician. Wife of Novikov 40
KHARITON, Anna Borisovna (Ania) (1901-94), sister of Iu.B. Khariton. Married name: Zakharovskaia 122
KHARITON Boris Osipovich (1877-1940), father of Iu.B. Khariton 122
KHARITON, Iulii Borisovich (1904-96), physicist, Academician. scientific director of Arzamas-16 1947-92 ix-x, xiii-xiv, xviii-xx, xxii, 35, 37, 43-4, 47, 50-3, 55, 57, 64-6, 68, 72, 76, 78-80, 84, 89, 96, 99, 105, 114-16, 119-20, 121-40, 142-3, 146, 148, 158-9, 163, 165
KHARITON, Lidia Borisovna (Lida) (1899-1974), sister of Iu. B. Khariton. Married name: Chernenko 122
KHARITON, Maria Nikolaevna (1902-77), wife of Iu.B. Khariton from 1929. Maiden name: Vul'fovich. Stage name: Zhukovskaia 111-12, 119-20, 138-9
KHARITON, Tat'iana Iul'evna (Tata) (1926-85), daughter of Iu.B. Khariton 143
KHLOPIN, Vitalii Grigor'evich (1890-1950), radiochemist xxv
KHRUSHCHEV, Nikita Sergeevich (1894-1971), Soviet political leader 89n., 114n.

KIENIA, Mikhail Grigor'evich, accountant 109
KIKOIN, Isaak Konstantinovich (1908-84), physicist 133
KIROV, Sergei Mironovich (1886-1934). Prominent Bolshevik, whose murder in December 1934 unleashed Stalinist terror 7
KLODT, Petr Karlovich, (1805-67), sculptor 13, 45
KOCHARIANTS, Samvel Grigor'evich, chief engineer at Arzamas-16 xvii, xix
KOMEL'KOV, Vladimir Stepanovich, physicist 57-8, 72
KONDRAT'EV, Viktor Nikolaevich (1902-79), physicist 123
KORMER, Samuil Borisovich (1922-82), physicist 53, 55, 59, 62, 68, 70, 98-101
KOSTAN'IAN, Kh.A. official at Arzamas-16 xiii
KOVALEVA, Marina Frantsevna, friend of V.A. Tsukerman and Z.M. Azarkh 15, 19
KOZHINA, Nina Konstantinovna, radiographer 9
KOZINTSEV, Grigorii Mikhailovich (1905-73), film director xv
KRAININ, pseudomym of Tsukerman and Azarkh xxi
KRUPNIKOV, Konstantin Konstantinovich (b. 1922) nuclear physicist 53, 55, 74, 78
KURCHATOV, Igor Vasil'evich (1903-60), physicist, Academician xii, xxiv-xxv, xxix-xxx, 47, 50, 52, 72, 75-6, 78, 82, 89-90, 92-5, 98, 109-10, 111-20, 121, 133, 134, 137-8, 140, 142, 146, 151, 154, 159
KURNOSOVA, Lidia Vasil'evna (Lida), wife of O.N. Vavilov 29, 31-2
KUZ'MINA, Nina Ivanovna, director of cultural centre 167
LANDAU, Lev Davydovich (1908-68), physicist. Nobel Prize for Physics 1962 43, 46, 72, 102, 130
LANGEVIN, Paul (1872-1946), French physicist 128
LAVRENT'EV, Mikhail A. (1900-80), mathematician, Academician 36, 154
LEDENEV, Boris Nikolaevich, physicist 53, 61, 78
LENIN, Vladimir Il'ich (1870-1924), founder of the Soviet state 17
LERMONTOV, Mikhail Iur'evich (1814-41), major poet 42
LESKOV, Nikolai Semenovich (1831-95), writer 107
LEVITAN, Iurii Borisovich (1914-), Soviet radio announcer 22
LIUDAEV, Robert Zakharovich, physicist xxii, 152
LOBACHEVSKII, Nikolai Ivanovich (1793-1856), mathematician 24
LOGONOVSKII Aleksandr Vasil'evich (1812[?]-55), sculptor 13
LOMINSKII, Georgii Pavlovich, explosives expert 60
LOPSCHITZ, A.M., mathematician 150
LYSENKO, Trofim Denisovich (1898-1976), agronomist 30, 83, 146n., 154-5
MACH, Ernst (1836-1916), Austrian physicist and philosopher 11, 146n.
MAIAKOVSKII, Vladimir Vladimirovich (1893-1930), poet and dramatist 101n., 139
MAKHNEV, V.A., security official xii
MAKOVSKII, Konstantin Egor'evich (1839-1915), painter 13
MALENKOV, Georgii Maksimilianovich (1902-88), Soviet politician. Prime Minister 1953-55 xi-xii, 89

Index of names

MAL'SKII, Anatolii Iakovlevich, factory director 68, 76
MANAKOVA, Maria Alekseevna (b. 1913), radiography specialist 53-5, 60, 65
MARCHENKO, Anatolii Tikhonovich (1938-86), dissident writer 158
MEITNER, Lise (1878-1968), Austrian physicist xxix, 131-2
MENDEL, Gregor (1822-84), Austrian botanist 83
MESHIK, Pavel V., security official. Executed 1953 84
METTER, Izrail Moiseevich (b. 1909), short story writer xv
MIROKHIN, Iurii Valentinovich, Arzamas-16 staff member 74
MODEL, Il'ia Shulimovich (d. 1996) physicist xxii, 52-3, 57, 74-5
MOKHOV, Vladislav Nikolaevich, staff member at Arzamas-16 xiii
MOLOTOV, Viacheslav Mikhailovich (1890-1986), Soviet Foreign Minister 1939-49, 1953-56 xvii, 22
MOROZ, Oleg, journalist xi
MOZART, Wolfgang Amadeus (1756-91), Austrian composer 119-20
MUSORGSKII, Modest Petrovich (1839-81), composer 104
MUZRUKOV, Boris Glebovich (1904-78), metallurgist, factory director 117-8, 163-8
NAGIBIN, Iurii Markovich (1920-94), Soviet writer x, xv, 18n.
NEGIN, Evgenii Arkad'evich, former director of Arzamas-16 xi
NODDACK[-TACKE], Ida (1896-1978), German chemist 131-2
NOVIKOV, Petr Sergeevich (1901-75), mathematician. Husband of Keldysh 40
NOVITSKII, Ananii Il'ich, scientist at Arzamas-16 87
NUZHDIN, Nikolai Ivanovich (b. 1904), biologist 154-5
OBREIMOV, Ivan Vasil'evich (1894-1981), physicist 43, 130
OGNEV, Leonid Ivanovich, staff member at Arzamas-16 xiii
OKUDZHAVA, Bulat Shalvovich (1924-97), poet and novelist xxxiii
ORLOV, F.I., official xiii
PASTERNAK, Boris (1890-1960), poet and writer xiv, 121
PAVLOVSKAIA, Nella Grigor'evna, physicist xxii, 103-4
PAVLOVSKII, Aleksandr Ivanovich, physicist 152
PEIERLS, Rudolf Ernest (1907-96), German-British theoretical physicist xxx, 132
PEREL'MAN, Iakov Isidorovich (1882-1942), popularizer of Science 122
PERVUKHIN, Mikhail Grigor'evich (1904-78), Soviet politician xii
PETROV, Nikolai Aleksandrovich, factory director xiii
PETROV, Nikolai Vasil'evich (1890-1964), theatre director 139
PETRZHAK, Konstantin Antonovich (b. 1908) nuclear physicist xxv, 137
PISARZHEVSKII, Lev Vladimirovich (1864-1938), chemist, Academician 83
PUSHKIN, Aleksandr Sergeevich (1799-1837), Russia's national poet xxxiv, 66, 84
RACHMANINOV, Sergei Vasil'evich (1873-1943), composer 139
RAEVSKII, Nikolai Petrovich, physicist 44
RAMAZANOV, Nikolai Aleksandrovich (1815-67), sculptor 13
REPIN Il'ia Efimovich (1844-1930), painter 81
ROSTROPOVICH, Mstislav Leopol'dovich (b. 1927), cellist xvi

ROZING, Vladimir S., assistant to Iu.B. Khariton 130
ROZOV, Viktor Sergeevich (b. 1913), playwright xv
RUBASHOVA, Khasia Zalmanovna (d. 1941), aunt of V.A. Tsukerman and Z.M. Azarkh xx
RUBASHOVA, Rika Zalmanovna (d. 1939), aunt of V.A. Tsukerman and Z.M. Azarkh xx
RUBASHOVA, Tsilia (Tseita) Zalmanovna (d. 1942), aunt of V.A. Tsukerman and Z.M. Azarkh xx
RUSINOV, Lev I., scientist xxix
RUTHERFORD, Ernest (1871-1937), New Zealand-born British chemist. Nobel Prize for Chemistry 1908. Knighted 1914. Became Baron Rutherford of Nelson 1931 126-9
SAINT-EXUPÉRY, Antoine de (1900-44), French writer xxi
SAINT-SAËNS, Camille (1835-1921), French composer 104
SAKHAROV, Andrei Dmitrievich (1921-89), physicist, campaigner for human rights x-xi, xvii-xx, xxiv, xxvi, xxx-xxxi, 54, 84, 90, 143, 149-58
SAKHAROV, Dmitrii Ivanovich (1899-1961), father of the above 150
SCHALL, Rudi, German scientist 40-1
SEMENOV, Nikolai Nikolaevich (1896-1986), physicist and chemist. Shared Nobel Prize for Chemistry 1956 xxv, 43, 71, 123-6
SEMIN, Anatolii Borisovich, doctor 69-70
SERAPHIM OF SAROV (1759-1833), saint of the Orthodox Church. Originally buried in the monastery at Sarov, his remains now lie at nearby Diveevo x, 49, 134
SEVAST'IANOV, Nikolai Grigor'evich, laboratory assistant 7
SHAL'NIKOV, Aleksandr Iosifovich (b. 1905) physicist 36, 43, 126, 130
SHATROV, Mikhail Filippovich (b. 1932), dramatist xxi
SHAVYRIN, Boris Ivanovich (1902-65), weapons designer 33-4
SHCHELKIN, Kirill Ivanovich (1911-68), physicist xviii, xxxi, 53, 76, 78
SHMIDT, Otto Iul'evich (1891-1956), mathematician, geophysicist, polar explorer, Academician 24-5
SHOROKH, Anatolii Andreevich (1926-87), Arzamas-16 staff member 74
SHVARTS, Evgenii L'vovich (1896-1958), dramatist 139
SIBELIUS, Jean (1865-1957), Finnish composer 139
SILKIN, Aleksandr Stepanovich, head of political department at Arzamas-16. 165
SIMONOV, Konstantin Mikhailovich (1915-79), poet and novelist xv
SINITSYN, Aleksandr Fedorovich (b. 1924), aeronautical engineer 7-8
SLAVSKII, Efim Pavlovich (1898-1991), Soviet government minister xviii
SLUTSKII, Boris Abramovich, (1919-86), poet 78
SMAGIN, Boris, scientist xi
SMYTH, Henry deWolf (1898-1987), physicist. Adviser to U.S. government 47
SNITKO, Konstantin Konstantinovich, army officer 35
SOBINOV, Leonid Vital'evich (1872-1934), lyric tenor 139
SOF'IN, Aleksandr Petrovich (d. 1942), artist, husband of Sof'ina 102
SOF'INA, Vera Viktorovna (1897-1975), explosives expert 53, 71, 102-4, 166

Index of names

SOKOLIK, I.A. (d. 1960), designer and inventor 57
SPERANSKAIA, Maria Parfen'evna (Marusia), wife of L.V. Al'tshuler 29, 32, 53
SPERANSKAIA, Tat'iana Parfen'evna (Tania), sister of the above 24
STALIN, Iosif Vissarionovich (1879-1953), Soviet dictator 1929-53 xvii, 44, 153n., 160
STANIUKOVICH, K.P., physicist 71, 102
STRASSMANN, Friedrich (Fritz) (1902-80), German chemist xxix
SURIKOV, Vasilii Ivanovich (1848-1916), painter 13
TAMM, Igor Evgen'evich (1895-1971), physicist. Nobel Prize for Physics 1958 xxiv, xxxi, 9, 43, 54 64, 145-8, 150-1, 154-5, 157
TARASOV, Diodor Mikhailovich (1911-1974), physicist 50, 53, 55, 62, 73-4, 78, 104-6
TARASOV, Mikhail Semenovich (b. 1917), radio engineer 57, 88-9
TATARKSII, Veniamin Vol'fovich (1917-76) expert in radiography 54
TCHAIKOVSKII, Petr Il'ich (1840-93), composer 13, 110
TELLER, Edward (b.1908), Hungarian-American nuclear scientist 78, 85
THOMSON, Joseph John (1856-1940), English physicist. Nobel Prize for Physics 1906. Knighted 1908 128
TIKHONOV, Nikolai Semenovich (1896-1979), poet 79
TIMIRIAZEV, Kliment Arkad'evich (1843-1920), botanist 24
TIUTCHEV, Fedor Ivanovich (1803-73), poet 90n.
TOCHILOVSKII, Pavel Mikhailovich, technician 106-7
TOLSTOI, Fedor Petrovich (1783-1873), painter 13
TOLSTOI, Lev Nikolaevich (1828-1910), novelist and thinker x
TRIFONOV, Iurii Valentinovich (1925-81), prose writer 1, 6
TRUMAN, Harry S. (1884-1972), U.S. President 1945-52 45-6, 67, 77
TRUTNEV, Iurii, Soviet scientist xii
TSIRKOV, Georgii A., physicist 152
TSUKERMAN, Aleksandr Veniaminovich (Sasha) (1949-66), son of V.A. Tsukerman x, 14-15
TSUKERMAN, Aron (d. 1922), father of V.A. Tsukerman xx, 3
TSUKERMAN, Irina Veniaminovna (b. 1937), daughter of V.A. Tsukerman x, xx, 3n., 12, 14, 21-2, 44, 49
TSUKERMAN, Kreina Zalmanovna (Katia; 1888-1944), mother of V.A.Tsukerman xx, 2-4, 14, 32, 41-3
TSUKERMAN, Samarii Aronovich (Shura; 1918-79), brother of V.A. Tsukerman 3, 4, 14, 42
TSUKERMAN, Veniamin Aronovich (1913-93) *passim*
TSVETAEVA, Marina Ivanovna (1892-1941), poet xxi
VAL'TA, Zinaida Frantsevna (1904-c.1950), physicist 125-6
VAL'TER, Aleksandr Filippovich (1898-1941), physicist 123
VANNIKOV, Boris L'vovich (1897-1962), politician xvii-xviii, 38, 72
VAVILOV, Nikolai Ivanovich (1887-1942), geneticist. Main victim of campaign led by T.D. Lysenko. Died in prison camp 43
VAVILOV, Oleg Nikolaevich (1920-1946), physicist. Son of N.I. Vavilov. Husband of L.V. Kurnosova 26, 29-31
VAVILOV, Sergei Ivanovich (1891-1951), physicist, Academician 43

VEKSLER, Vladimir Iosifovich (1907-66), physicist 26, 30. 36
VENTSEL, Dmitrii Aleksandrovich (b.1898), aeronautical engineer 141
VERESHCHAGIN, Vasilii Vasil'evich (1842-1904), painter 13
VERNADSKII, Vladimir Ivanovich (1863-1945), mineralogist and geochemist, Academician 121
VISHNEVSKAIA, Galina Pavlovna (b. 1926), wife of Rostropovich xvi
VOINOV, Aleksei Mikhailovich (b. 1927), friend of M.V. Dmitriev xxii, 97
VOINOVA, Galina Borisovna, friend of M.V. Dmitriev. Wife of A.M. Voinov xxii, 97
VOROSHILOV, Kliment Efremovich (1881-1969), Marshal of the Soviet Union. Titular Head of State 1953-60 22
VOZNESENSKII, Andrei Andreevich, poet xx, 41
VOZNESENSKII, Nikolai Alekseevich (1903-50), official xii
VYSOTSKII, Vladimir Semenovich (1938-80), poet and songwriter 30
WAGNER, Richard (1813-83), German composer 139
WALTON, Ernest Thomas Sinton (b. 1903), Irish physicist 112n.
WATSON, James (b. 1928). American biologist. Nobel Prize for Medecine 1962 (with F. Crick and Maurice Wilkins) 146
ZABABAKHIN, Evgenii Ivanovich (1917-88), physicist, Academician xxxi, 54, 84, 105. 116, 141
ZAKHARENKOV, Aleksandr Dmitrievich (1921-89), Soviet politician. Deputy Minister of Medium Machine Building 50
ZAKHAROV, Aleksandr, scientist xvi
ZAKHAROVA, Tat'iana Vasil'evna, expert in radiography 52-4, 115
ZAVENIAGIN, Avraamii Pavlovich (1901-56), Minister of Medium Machine Building 1955-56 xii, xvii-xvii, 75, 84, 133-4
ZAVOISKII, Evgenii Konstantinovich (1907-76), physicist 72
ZELENAIA, Rina Vasil'evna (b. 1902), actress 139
ZELENSKII, Konstantin Fedorovich, physicist 56
ZEL'DOVICH, Iakov Borisovich (1914-1987), nuclear physicist, Academician xvii, xxiv-xxv, xxix, xxxi, 35, 37, 43, 51, 53-4, 64, 68, 72-5, 78, 83-5, 89-90, 104-5, 112, 116, 126, 130-3, 137, 140-4, 149-50
ZERNOV, Pavel Mikhailovich (1905-64), director of Arzamas-16 1946-51 xi, xviii, xxxi, 50, 58-9, 66, 68, 76-8, 107-9, 134, 158-63
ZHURAVLEV, Aleksandr Alekseevich, glassblower 110
ZINCHENKO, Lev, driver 74
ZLATOUSTOVSKII, V.B., soldier xii
ZYKOV, Anatolii Petrovich, Arzamas-16 staff member 74